Compression in Video and Audio

John Watkinson

Focal Press
An imprint of Butterworth-Heinemann Ltd
Linacre House, Jordan Hill, Oxford OX2 8DP

A member of the Reed Elsevier plc group

OXFORD LONDON BOSTON
MUNICH NEW DELHI SINGAPORE SYDNEY
TOKYO TORONTO WELLINGTON

First published 1995
Reprinted 1995

© John Watkinson 1995

All rights reserved. No part of this publication
may be reproduced in any material form (including
photocopying or storing in any medium by electronic
means and whether or not transiently or incidentally
to some other use of this publication) without the
written permission of the copyright holder except in
accordance with the provisions of the Copyright,
Designs and Patents Act 1988 or under the terms of a
licence issued by the Copyright Licensing Agency Ltd,
90 Tottenham Court Road, London, England W1P 9HE.
Applications for the copyright holder's written permission
to reproduce any part of this publication should be addressed
to the publishers

British Library Cataloguing in Publication Data
A catalogue record for this book is available from
the British Library

Library of Congress Cataloguing in Publication Data
A catalogue record for this book is available from
the Library of Congress

ISBN 0 240 51394 0

Composition by Genesis Typesetting, Rochester, Kent
Printed and bound in Great Britain by Clays Ltd, St Ives plc

For Howard and Matthew

Contents

Preface		*xi*
Acknowledgements		*xii*

Chapter 1 Introduction to compression — 1

1.1	Why compression is necessary	1
1.2	Some applications of compression	1
1.3	Lossless and perceptive coding	3
1.4	Compression principles	4
1.5	Drawbacks of compression	7
1.6	Audio compression	8
1.7	Sub-band coding	8
1.8	Transform coding	8
1.9	Predictive coding	8
1.10	Video compression	9
1.11	Intra-coded compression	9
1.12	Inter-coded compression	10
1.13	Introduction to motion compensation	11
1.14	Film-originated video compression	13
1.15	Some guidelines	15
	Reference	15

Chapter 2 Fundamentals — 16

2.1	What is an audio signal?	16
2.2	What is a video signal?	16
2.3	Types of video	17
2.4	What is a digital signal?	18
2.5	Introduction to conversion	21
2.6	Sampling and aliasing	22
2.7	Reconstruction	27
2.8	Filter design	27
2.9	Sampling clock jitter	32
2.10	Choice of audio sampling rate	33
2.11	Video sampling structures	35

2.12	The phase-locked loop		37
2.13	Quantizing		38
2.14	Quantizing error		39
2.15	Dither		43
2.16	Binary codes for audio		45
2.17	Binary codes for component video		50
2.18	Introduction to digital processes		51
2.19	Logic elements		52
2.20	Storage elements		54
2.21	Binary adding		57
2.22	Gain control by multiplication		57
	References		59

Chapter 3 Processing for compression — 61

3.1	Filters	61
3.2	The quadrature mirror filter	67
3.3	Filtering for video noise reduction	72
3.4	Transforms	73
3.5	The Fourier transform	77
3.6	The discrete cosine transform (DCT)	86
3.7	The wavelet transform	89
3.8	Motion compensation	91
3.9	Motion estimation techniques	91
3.10	Compression and requantizing	98
	References	102

Chapter 4 Audio compression — 103

4.1	Psychoacoustics and masking	103
4.2	Codec level calibration	104
4.3	Quality measurement	105
4.4	The limits	106
4.5	Compression applications	107
4.6	Audio compression techniques	108
4.7	Non-uniform coding	109
4.8	Floating-point coding	110
4.9	Predictive coding	112
4.10	Sub-band coding	114
4.11	Transform coding	115
4.12	A simple sub-band coder	117
4.13	Data reduction formats	119
4.14	ISO Layer I – simplified MUSICAM	120
4.15	ISO Layer II – MUSICAM	120
4.16	ISO Layer III	121
4.17	apt-x100	121
4.18	Dolby AC-2	122
4.19	PASC	122
4.20	The ATRAC coder	125
	References	126

Chapter 5 Video compression   **128**

5.1	The eye	128
5.2	Colour vision	132
5.3	Colour difference signals	134
5.4	Motion and resolution	136
5.5	Applications of video compression	139
5.6	Intra-coded compression	141
5.7	JPEG compression	146
5.8	Compression in Digital Betacam	148
5.9	Inter-coded compression	149
5.10	Error propagation	152
5.11	CCIR Rec. 723 compression	155
5.12	Introduction to MPEG coding	157
5.13	MPEG-1 coding	160
5.14	MPEG-2 coding	164
5.15	Coding artifacts	168
	References	168

Index   171

Preface

Compression technology has been employed for a long time, but until recently the technology was too complex for everyday applications. The onward march of LSI technology means that increasingly complex processes become available at moderate cost. Compression has now reached the stage where it can economically be applied to video and audio systems on a wide scale.

This book recognizes the wide application of compression by treating the subject from first principles without assuming any particular background for the reader. Whilst the subject is traditionally described mathematically, it is the author's view that mathematics is no more than a technical form of shorthand *describing* a process. As such it cannot *explain* anything and has no place in a book of this kind where the majority of the explanations are in plain English.

An introductory chapter is included which suggests some applications of compression and how it works in a simplified form. Additionally a fundamentals chapter contains all of the background necessary to follow the rest of the book.

Theory is balanced with a wide range of practical applications in transmission and recording. It will be seen that the compression techniques which can be used are restricted by the application and must be carefully tailored to it.

Throughout this book the reader will find notes of caution and outlines of various pitfalls for the unwary. Compression is a useful tool in certain applications but it is a dangerous master and if used indiscriminately the results are bound to be disappointing. Various descriptions of kinds of artifacts and impairments which result from the misuse of compression are included here. Compression is rather like a box of fireworks: used wisely a pleasing display results – on the other hand it can blow up in your face.

To suggest that compression should never be used to ensure that disasters are avoided is as naive as trying to ban fireworks. The solution is the same: to explain the dangers and propose safe practices. Retire to a safe place and read this book.

John Watkinson
Burghfield Common 1995

Acknowledgements

Information for this book has come from a range of sources to whom I am indebted. I would specifically mention the publications of the ISO, the AES and SMPTE which provided essential groundwork. I have had numerous discussions with Peter de With of Philips, Steve Lyman of CBC, Bruce Devlin and Mike Knee of Snell and Wilcox and Peter Kraniauskas which have all been extremely useful.

Chapter 1
Introduction to compression

1.1 Why compression is necessary

Compression, bit rate reduction and data reduction are all terms which mean basically the same thing in this context. In essence the same information is carried using a smaller quantity or rate of data. It should be pointed out that in audio parlance, *compression* traditionally means a process in which the dynamic range of the sound is reduced. In digital parlance the same word means that the bit rate is reduced, ideally leaving the dynamics of the signal unchanged. Provided the context is clear, the two meanings can co-exist without a great deal of confusion.

There are two fundamental reasons why compression techniques are used:

(a) To make possible some process which would be impracticable without it.
(b) To perform a known process more economically.

In the case of Video-CD the challenge was to produce moving video from an almost standard Compact Disc player. The data rate of a Compact Disc is such that recording of conventional uncompressed video would be out of the question and so Video-CD represents an example of the first category.

In the case of Video-On-Demand, technology exists to convey full bandwidth video to the home, but to do so for a single individual would be prohibitively expensive. Compression does not make Video-On-Demand possible, it makes it economically viable.

1.2 Some applications of compression

Compression is summarized in Figure 1.1. It will be seen in (a) that the data rate is reduced at source by the *compressor*. The compressed data are then passed through a communication channel and returned to the original rate by the *expander*. The ratio between the source data rate and the channel data rate is called the *compression factor*. The term *coding gain* is also used. Sometimes a compressor and expander in series are referred to as a *compander*. The compressor may equally well be referred to as a *coder* and the expander a *decoder* in which case the tandem pair may be called a *codec*. In communications, the cost of data links is often roughly proportional to the data rate and so there is simple economic pressure to use a high compression factor. However, it

1

2 Introduction to compression

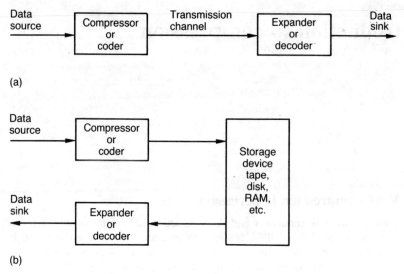

Figure 1.1 In (a) a compression system consists of a compressor or coder, a transmission channel and a matching expander or decoder. The combination of coder and decoder is known as a codec. In (b) compression can be used around a recording medium. The storage capacity may be increased or the access time reduced according to the application.

should be borne in mind that implementing the codec also has a cost which rises with compression factor and so it will be necessary to compromise.

Figure 1.1(b) shows the use of a codec with a recorder. The use of compression in recording applications is extremely powerful. The playing time of the medium is extended in proportion to the compression factor. In the case of tapes, the access time is improved because the length of tape needed for a given recording is reduced and so it can be rewound more quickly.

In DAB (Digital Audio Broadcasting), also known as DR (Digital Radio), and in digital television transmission, compression is used to reduce the bandwidth needed. There is only one electromagnetic spectrum and pressure from other services makes efficient use of bandwidth mandatory. Fortunately in broadcasting there is a mass market for decoders and these can be implemented as low-cost integrated circuits. Fewer encoders are needed and so it is less important if these are expensive.

In DCC (Digital Compact Cassette) 4:1 data reduction is used on the audio data prior to recording. This means that the same tape speed as for analog cassettes can be used without excessively short wavelengths or narrow tracks being needed on the tape. As a result, conventional chrome cassette tape can be used instead of the more expensive metal tapes. The transport design is less critical and more contamination can be tolerated.

In workstations designed for the editing of audio and/or video, the source material is stored on hard disks for rapid access. Today's disk drives cannot offer sufficient storage capacity to give realistic playing times without the use of compression. When a workstation is used for *off-line* editing, a high compression factor can be used and artifacts will be visible in the picture. This is of no consequence as the picture is only seen by the editor who uses it to make an EDL (Edit Decision List) which is no more than a list of actions and the timecodes at which they occur. The original uncompressed material is then *conformed* to the EDL to obtain a high quality edited work. When *on-line* editing is being performed, the output of the workstation is the finished product and clearly a lower compression factor will have to be used.

In the Digital Betacam DVTR it was a goal that the machine should use the same cassette shell as the analog Betacam recorders so that some backwards compatibility would be possible. In order to obtain a reasonable playing time in the confines of the existing cassette it was necessary to adopt a mild compression stage. In the Ampex DCT (Digital Component Technology) DVTR the goal was to record component digital video on a transport originally designed for composite digital which needs only half the bit rate. Once more a mild compression stage was used.

1.3 Lossless and perceptive coding

Although there are many different coding techniques, all of them fall into one or other of these categories. In *lossless* coding, the data from the expander are identical bit-for-bit with the original source data. The Stacker programs which increase the apparent capacity of disk drives in personal computers use lossless codecs. Clearly with computer programs the corruption of a single bit can be catastrophic. Lossless coding is generally restricted to compression factors of around 2:1. It is important to appreciate that a lossless coder cannot guarantee a particular compression factor and the communications link or recorder used with it must be able to function with the variable output data rate. Source data which result in poor compression factors on a given codec are known as *difficult* material. It should be pointed out that the difficulty is often a function of the codec. In other words, data which one codec finds difficult may not be found difficult by another. Lossless codecs can be included in bit-error-rate testing schemes. It is also possible to cascade or *tandem* lossless codecs without any special precautions.

In *lossy* coding, data from the expander are not identical bit-for-bit with the source data and as a result comparing the input with the output is bound to reveal differences. Lossy codecs are not suitable for computer data, but are popular for audio and video applications as they allow greater compression factors than lossless codecs. Successful lossy codecs are those in which the differences are arranged so that a human viewer or listener finds them subjectively difficult to detect. Thus lossy codecs must be based on an understanding of psychoacoustic and psychovisual perception and are often called *perceptive* codes. The generation of colour difference signals from RGB in video represents an application of perceptive coding. The human viewer sees no change in quality although the bandwidth of the colour difference signals is reduced.

In perceptive coding, the greater the compression factor required, the more accurately must the human senses be modelled. Perceptive coders can be forced to operate at a fixed compression factor. This is convenient for practical recording and transmission applications where a fixed data rate is easier to handle than a variable rate. The result of a fixed compression factor is that the subjective quality can vary with the 'difficulty' of the input material. Perceptive codecs should not be connected in tandem indiscriminately, especially if they use different algorithms. As the reconstructed signal from a perceptive codec is not bit-for-bit accurate, clearly such a codec cannot be included in any bit-error-rate testing system as the coding differences would be indistinguishable from real errors.

1.4 Compression principles

In a PCM digital system the bit rate is the product of the sampling rate and the number of bits in each sample and this is generally constant. The *information* rate of a real signal varies, however. The difference between the information rate and the bit rate is known as the redundancy. Compression systems are designed to eliminate that redundancy. One way in which this can be done is to exploit statistical predictability in signals. The information content or *entropy* of a sample is a function of how different it is from the predicted value. Most signals have some degree of predictability. A sine wave is highly predictable, because all cycles look the same. According to Shannon's theory, any signal which is totally predictable carries no information. In the case of the sine wave this is clear because it represents a single frequency and so has no bandwidth.

At the opposite extreme, a signal such as noise is completely unpredictable and as a result all codecs find noise *difficult*. There are two consequences of this characteristic. Firstly, a codec which is designed using the statistics of real material should not be tested with random noise because it is not a representative test. Secondly, a codec which performs well with clean source material may perform badly with source material containing superimposed noise. Most practical compression units require some form of pre-processing before the compression stage proper, and appropriate noise reduction should be incorporated in the pre-processing if noisy signals are anticipated. It will also be necessary to restrict the degree of compression applied to noisy signals.

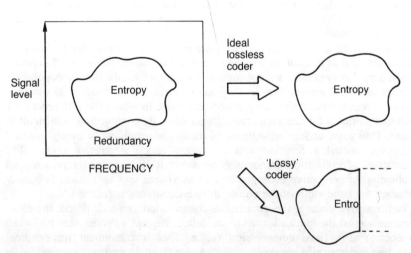

Figure 1.2 A perfect coder removes only the redundancy from the input signal and results in subjectively lossless coding. If the remaining entropy is beyond the capacity of the channel some of it must be lost and the codec will then be lossy. An imperfect coder will also be lossy as it fails to keep all entropy.

All real signals fall part-way between the extremes of total predictability and total unpredictability or noisiness. If the bandwidth (set by the sampling rate) and the dynamic range (set by the wordlength) of the transmission system are used to denote an area, this sets a limit on the information capacity of the channel. Figure 1.2 shows that most real signals only occupy part of that area. The signal may not contain all frequencies, or it may not have full dynamics at certain frequencies. The entropy is the actual area occupied by the signal. This is the area that *must* be transmitted if there are to be no subjective differences or *artifacts* in the received signal. The remaining area is called the *redundancy* because it adds nothing to the information conveyed. Thus an ideal coder could be imagined which miraculously sorts out the entropy from the redundancy and only sends the former. An ideal decoder would then recreate the original impression of the information quite perfectly. As the ideal is approached, the coder complexity rises. Furthermore, we would have to provide a channel which could accept whatever entropy the coder extracts in order to have transparent quality. If the channel capacity was not sufficient for that, then the coder would also have to discard some of the entropy and with it useful information. As a result, moderate coding gains which only remove redundancy need not cause artifacts and result in systems which are described as *subjectively lossless*. Larger coding gains which remove the redundancy as well as some of the entropy must result in artifacts. It will also be seen from Figure 1.2 that an imperfect coder will fail to separate the redundancy and may discard entropy instead, resulting in artifacts at a sub-optimal compression factor.

A variable rate transmission or recording channel is inconvenient and unpopular with channel providers because it is difficult to police. The requirement can be overcome by combining several compressed channels into one constant rate transmission in a way which allocates data rate flexibly between the channels. The probability of all channels reaching peak entropy at once is very small and so those channels which are at one instant passing easy material will free up transmission capacity for those channels which are handling difficult material.

Where the same type of source material is used consistently, e.g. English text, then it is possible to perform a statistical analysis on the frequency with which particular letters are used. Variable-length coding is used in which frequently used letters are allocated short codes and letters which occur infrequently are allocated long codes. This results in a lossless code. The well known Morse code used for telegraphy is an example of this approach. The letter 'e' is the most frequent in English and is sent with a single dot. An infrequent letter such as 'z' is allocated a long, complex pattern. It should be clear that codes of this kind which rely on a prior knowledge of the statistics of the signal are only effective with signals actually having those statistics. If Morse code is used with Polish, the transmission becomes significantly less efficient because the statistics are quite different; the letter 'z', for example, is quite common in Polish.

The Huffman code[1] is one which is designed for use with a data source having known statistics and shares the same principles with the Morse code. The probability of the different code values to be transmitted is studied, and the most frequent codes are arranged to be transmitted with short wordlength symbols. As the probability of a code value falls, it will be allocated longer wordlength. The Huffman code is used in conjunction with a number of compression techniques and is shown in Figure 1.3. The input or *source* codes are assembled in order of

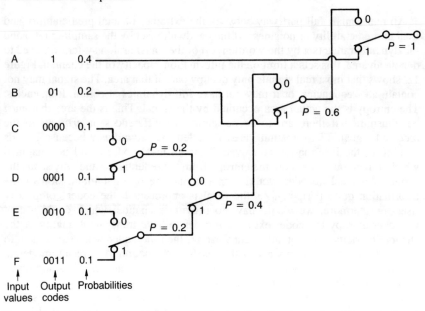

Figure 1.3 The Huffman code achieves compression by allocating short codes to frequent values. To aid deserializing the short codes are not prefixes of longer codes.

descending probability. The two lowest probabilities are distinguished by a single code bit and their probabilities are combined. The process of combining probabilities is continued until unity is reached and at each stage a bit is used to distinguish the path. The bit will be a 0 for the most probable path and 1 for the least. The compressed output is obtained by reading the bits which describe which path to take going from right to left.

In the case of computer data, there is no control over the data statistics. Data to be recorded could be instructions, images, tables, text files and so on; each having their own code value distributions. In this case a coder relying on fixed source statistics will be completely inadequate. Instead a system is used which can learn the statistics as it goes along. The Lempel–Ziv–Wekh (LZW) lossless codes are in this category. These codes build up a conversion table between frequent long source data strings and short transmitted data codes at both coder and decoder; initially their compression factor is below unity as the contents of the conversion tables are transmitted along with the data. However, once the tables are established, the coding gain more than compensates for the initial loss. In some applications, a continuous analysis of the frequency of code selection is made and if a data string in the table is no longer being used with sufficient frequency it can be deselected and a more common string substituted.

Lossless codes are less common for audio and video coding where perceptive codes are permissible. The perceptive codes often obtain a coding gain by shortening the wordlength of the data representing the signal waveform. This must increase the noise level and the trick is to ensure that the resultant noise is placed at frequencies where human senses are least able to perceive it. As a result, although the received signal is measurably different from the source data, it can *appear* the same to the human listener or viewer at moderate compression factors. As these codes rely on the characteristics of human sight and hearing, they can only be fully tested subjectively.

The compression factor of such codes can be set at will by choosing the wordlength of the compressed data. Whilst mild compression will be undetectable, with greater compression factors, artifacts become noticeable. Figure 1.2 shows that this is inevitable from entropy considerations.

1.5 Drawbacks of compression

By definition, compression removes redundancy from signals. Redundancy is, however, essential to making data resistant to errors. As a result, it is generally true that compressed data are more sensitive to errors than uncompressed data. Thus transmission systems using compressed data must incorporate more powerful error correction strategies and avoid compression techniques which are notoriously sensitive. As an example, the Digital Betacam format uses relatively mild compression and yet requires 20 per cent redundancy whereas the D-5 format does not use compression and only requires 17 per cent redundancy even though it has a recording density 30 per cent higher.

Techniques using tables such as the Lempel–Ziv–Wekh codes are very sensitive to bit errors as an error in the transmission of a table value results in bit errors every time that table location is accessed. This is known as error propagation. Variable-length techniques such as the Huffman code are also sensitive to bit errors. As there is no fixed symbol size, the only way the decoder can parse a serial bit stream into symbols is to increase the assumed wordlength a bit at a time until a code value is recognized. The next bit must then be the first bit in the next symbol. A single bit in error could cause the length of a code to be wrongly assessed and then all subsequent codes would also be wrongly decoded until synchronization could be re-established. Later variable length codes sacrifice some compression efficiency in order to offer better resynchronization properties.

In non-real-time systems such as computers an uncorrectable error results in reference to the back-up media. In real-time systems such as audio and video this is impossible and concealment must be used. However, concealment relies on redundancy and compression reduces the degree of redundancy. Media such as hard disks can be verified so that uncorrectable errors are virtually eliminated, but tape is prone to dropouts which will exceed the burst-correcting power of the replay system from time to time. For this reason the compression factors used on audio or video tape should be moderate.

As perceptive coders introduce noise, it will be clear that in a cascaded system the second codec could be confused by the noise due to the first. If the codecs are identical then each may well make the same decisions when they are in tandem, but if the codecs are not identical the results could be disappointing. Signal manipulation between codecs can also result in artifacts which were previously undetectable being revealed because the signal which was masking them is no longer present. In general, compression should not be used for its own sake, but only where a genuine bandwidth or cost bottleneck exists. Even then the mildest compression possible should be used. Whilst high compression factors are permissible for final delivery of material to the consumer, they are not advisable prior to any post-production stages.

One practical drawback of compression systems is that they are largely generic in structure and the same hardware can be operated at a variety of compression factors. Clearly the higher the compression factor, the cheaper the system will be to operate so there will be economic pressure to use high compression factors. Naturally the risk of artifacts is increased and so there is counter-pressure from those with engineering skills to moderate the compression. The way of the world at the time of writing is that the accountants have the upper hand. This was not a problem when there were fixed standards such as PAL and NTSC, as there was no alternative but to adhere to them. Today there is a danger that the variable compression factor control will be turned too far in the direction of economy.

1.6 Audio compression

Perceptive coding in audio relies on the principle of auditory masking, which is treated in detail in Section 4.1. Masking causes the ear/brain combination to be less sensitive to sound at one frequency in the presence of another at a nearby frequency. If a first tone is present in the input, then it will mask signals of lower level at nearby frequencies. The quantizing of the first tone and of further tones at those frequencies can be made coarser. Fewer bits are needed and a coding gain results. The increased quantizing error is masked by the presence of the first tone.

1.7 Sub-band coding

Sub-band coding mimics the frequency analysis mechanism of the ear and splits the audio spectrum into a large number of different bands. Signals in these bands can then be quantized independently. The quantizing error which results is confined to the frequency limits of the band and so it can be arranged to be masked by the programme material. The ISO standards which will be used in DAB are based on sub-band coding, as are those used in DCC (Digital Compact Cassette).

1.8 Transform coding

In transform coding the time-domain audio waveform is converted into a frequency domain representation such as a Fourier, Discrete Cosine or Wavelet Transform (see Chapter 3). Transform coding takes advantage of the fact that the amplitude or envelope of an audio signal changes relatively slowly and so the coefficients of the transform can be transmitted relatively infrequently. Clearly such an approach breaks down in the presence of transients, and adaptive systems are required in practice. Transients cause the coefficients to be updated frequently whereas in stationary parts of the signal such as sustained notes the update rate can be reduced. Discrete Cosine Transform (DCT) coding is used in the compression system of the Sony MiniDisc.

1.9 Predictive coding

In a predictive coder there are two identical predictors, one in the coder and one in the decoder. Their job is to examine a run of previous sample code values and to extrapolate forward to estimate or predict what the next code value will be.

This is subtracted from the *actual* next code value at the encoder to produce a prediction error which is transmitted. The decoder then adds the prediction error to its own prediction to obtain the output code value again.

1.10 Video compression

Video signals exist in four dimensions: the magnitude of the sample, the horizontal and vertical spatial axes and the time axis. Compression can be applied in any or all of those four dimensions. Video compression is generally divided into two basic categories. When individual pictures are compressed without reference to any other pictures, the time axis does not enter the process, which is therefore described as *intra-coded* (intra = within) compression. It is an advantage of intra-coded video that there is no restriction to the editing which can be carried out on the picture sequence. As a result production VTRs such as the Ampex DCT and Sony's Digital Betacam use intra-field compression. Cut editing may take place on the compressed data directly if necessary. As intra-coding treats each picture independently, it can employ certain techniques developed for the compression of still pictures. The ISO JPEG (Joint Photographic Experts Group) compression standards are in this category.

Greater compression factors can be obtained by taking account of the redundancy from one picture to the next. This involves the time axis and the process is known as *inter-coded* (inter = between) compression. The ISO MPEG (Moving Picture Experts Group) compression standards are in this category. As an individual picture may exist only in terms of the differences from a previous picture, editing must be undertaken with caution. Clearly editing cannot arbitrarily be performed on the compressed data stream as if a previous picture is removed by an edit, the difference data will be insufficient to recreate the current picture.

1.11 Intra-coded compression

Intra-coding works in three dimensions on the horizontal and vertical spatial axes and on the sample values. Analysis of typical television pictures reveals that whilst there is a high spatial frequency content due to detailed areas of the picture, there is a relatively small amount of energy at such frequencies. Often pictures contain sizeable areas in which the same or similar pixel values exist. This gives rise to low spatial frequencies. The average brightness of the picture results in a substantial zero frequency component. Simply omitting the high-frequency components is unacceptable as this causes an obvious softening of the picture. A coding gain can be obtained by taking advantage of the fact that the amplitude of the spatial components falls with frequency. It is also possible to take advantage of the eye's reduced sensitivity to noise in high spatial frequencies. If the spatial spectrum is divided into frequency bands the high-frequency bands can be described by fewer bits not only because their amplitudes are smaller but also because more noise can be tolerated. The Discrete Cosine Transform and the Discrete Wavelet Transform are both ways in which two-dimensional pictures can be described in the frequency domain; these are discussed in Chapter 3.

1.12 Inter-coded compression

Inter-coding takes further advantage of the similarities between successive pictures in real material. Instead of sending information for each picture separately, inter-coders will send the difference between the previous picture and the current picture in a form of differential coding. Figure 1.4 shows the principle. A picture store is required at the coder to allow comparison to be made between successive pictures and a similar store is required at the decoder to make the previous picture available. The difference data may be treated as a picture itself and subjected to some form of transform-based compression.

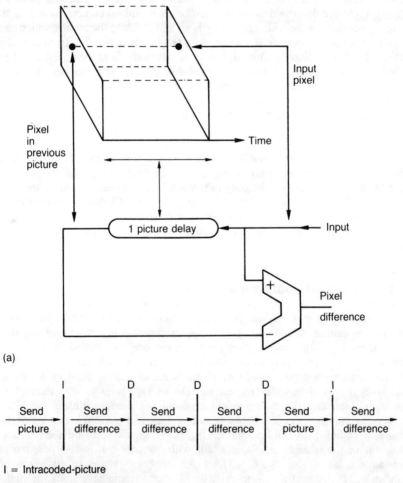

I = Intracoded-picture

D = Differentially coded picture

Figure 1.4 An inter-coded system (a) uses a delay to calculate the pixel differences between successive pictures. To prevent error propagation, intra-coded pictures (b) may be used periodically.

The simple system of Figure 1.4(a) is of limited use as, in the case of a transmission error, every subsequent picture would be affected. Channel switching in a television set would also be impossible. In practical systems a modification is required. One approach is the so-called 'leaky predictor' in which the next picture is predicted from a limited number of previous pictures rather than from an indefinite number. As a result errors cannot propagate indefinitely. The approach used in MPEG is that periodically some absolute picture data are transmitted in place of difference data. Figure 1.4(b) shows that absolute picture data, known as I or *intra* pictures are interleaved with pictures which are created using difference data, known as P or *predicted* pictures. The I pictures require a large amount of data, whereas the P pictures require less data. As a result the data rate varies dramatically and buffering has to be used to allow a constant transmission rate. The leaky predictor needs less buffering as the compression factor does not change so much from picture to picture.

It is advantageous from the standpoint of compression that a large number of P pictures should be present between I pictures. Conversely with such a structure the compressed bit stream can only be edited at I pictures as shown.

In the case of moving objects, although their appearance may not change greatly from picture to picture, the data representing them on a fixed sampling grid will change and so large differences will be generated between successive pictures. It is a great advantage if the effect of motion can be removed from difference data so that they only reflect the changes in appearance of a moving object since a much greater coding gain can then be obtained. This is the objective of motion compensation.

1.13 Introduction to motion compensation

In real television programme material objects move around before a fixed camera or the camera itself moves. Motion compensation is a process which effectively measures motion of objects from one picture to the next so that it can allow for that motion when looking for redundancy between pictures. Figure 1.5 shows that moving pictures can be expressed in a three-dimensional space which results from the screen area moving along the time axis. In the case of still objects, the only motion is along the time axis. However, when an object moves, it does so along the *optic flow axis* which is not parallel to the time axis. The optic flow axis joins the same point on a moving object as it takes on various screen positions.

It will be clear that the data values representing a moving object change with respect to the time axis. However, looking along the optic flow axis the appearance of an object only changes if it deforms, moves into shadow or rotates. For simple translational motions the data representing an object are highly redundant with respect to the optic flow axis. Thus if the optic flow axis can be located, a large coding gain can be obtained.

Figure 1.6 shows one way in which this concept can be considered. A delay allows two successive pictures to be compared in a motion estimator. This outputs *motion vectors* which are two-dimensional parameters describing the distance and direction by which certain parts of the picture moved between pictures. If the two pictures shown are superimposed, it will be seen that there is only redundancy or similarity in the stationary background. However, if one of the pictures is shifted according to the calculated motion vectors prior to

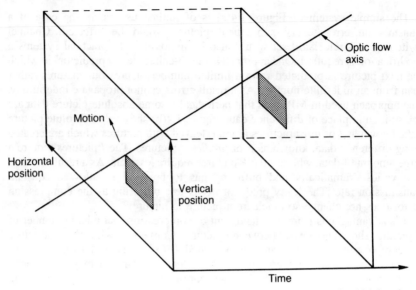

Figure 1.5 Objects travel in a three-dimensional space along the optic flow axis which is only parallel to the time axis if there is no movement.

superimposition, it will be found that there is now a great deal of similarity in the data representing the moving object.

A motion compensated coder works as follows. An *I* picture is sent, but is also locally stored so that it can be compared with the next input picture to find motion vectors for various areas of the picture. The *I* picture is then shifted according to these vectors to cancel inter picture motion, and compared with the next picture to produce difference data. It will be seen from Figure 1.7 that the difference data and the motion vectors are transmitted. At the receiver the original *I* picture is also held in a memory. It is also shifted according to the transmitted motion vectors and then the difference data are added to it to create

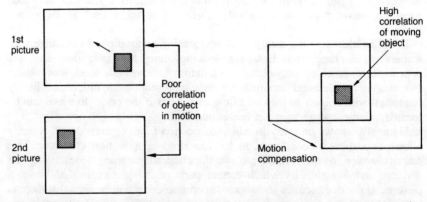

Figure 1.6 In motion compensation shifting a picture according to motion vectors cancels motion and reveals similarities from one picture to the next in moving objects.

Introduction to compression 13

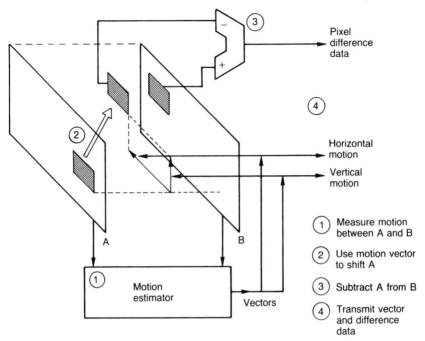

Figure 1.7 A motion-compensated compression system. The coder calculates motion vectors which are transmitted as well as being used locally to create a predicted picture. The difference between the predicted picture and the actual picture is transmitted.

a *P* picture. Any desired number of *P* pictures can be sent as motion vectors and difference data with respect to the original *I* picture. When sufficient *P* pictures have been transmitted, another *I* picture is sent and the process repeats. Clearly the difference data can be treated as pictures and can be subject to further compression techniques. The MPEG compression standards are based on processes of this kind.

1.14 Film-originated video compression

Film can be used as the source of video signals if a telecine machine is used. The most common frame rate for film is 24 Hz, whereas the field rates of television are 50 Hz and 60 Hz. This incompatibility is overcome in two different ways. In 50 Hz telecine, the film is simply played slightly too fast so that the frame rate becomes 25 Hz. Then each frame is converted into two television fields giving the correct 50 Hz field rate. In 60 Hz telecine, the film travels at the correct speed, but alternate frames are used to produce two fields then three fields. The technique is known as 3:2 pulldown. In this way two frames produce five fields and so the correct 60 Hz field rate results. The motion portrayal of 3:2 is not very good as moving objects judder with a fundamental period of five fields. Figure 1.8 shows how the optic flow is portrayed in film-originated video.

When film-originated video is input to a compression system, the disturbed optic flow will play havoc with the motion compensation system. In a 50 Hz system there appears to be no motion between the two fields which have

14 Introduction to compression

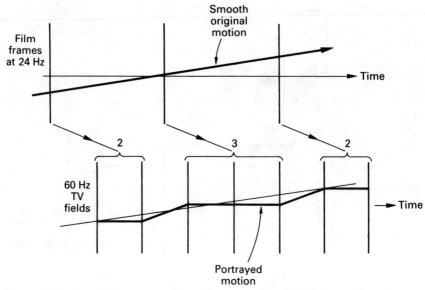

Figure 1.8 Telecine machines must use 3:2 pulldown to produce 60 Hz field rate video.

originated from the same film frame, whereas between the next two fields large motions will exist. In 60 Hz systems, the motion will be zero for three fields out of five.

With such inputs, it is more efficient to adopt a different processing mode which is based upon the characteristics of the original film. Instead of attempting to manipulate fields of video, the system de-interlaces pairs of fields in order to reconstruct the original film frames. This can be done by a fairly simple motion detector. When substantial motion is measured between successive fields in the output of a telecine, this is taken to mean that the fields have come from different film frames. When negligible motion is detected between fields, this is taken to indicate that the fields have come from the same film frame. In 50 Hz video it is quite simple to find the sequence and produce de-interlaced frames at 25 Hz. In 60 Hz 3:2 pulldown video the problem is slightly more complex because it is necessary to locate the frames in which three fields are output so that the third field can be discarded, leaving, once more, de-interlaced frames at 25 Hz. Whilst it is relatively straightforward to lock-on to the 3:2 sequence with direct telecine output signals, if the telecine material has been edited on videotape the 3:2 sequence may contain discontinuities. In this case it is necessary to provide a number of field stores in the de-interlace unit so that a series of fields can be examined to locate the edits. Once telecine video has been de-interlaced back to frames, intra- and inter-coded compression can be employed using frame-based motion compensation.

Compression schemes may include flags which teach the decoder the origin of the material. Material originating at 24 Hz but converted to interlaced video does not have the motion attributes of interlace because the lines in two fields have come from the same point on the time axis. Two fields can be combined to create a progressively scanned frame. In the case of 3:2 pulldown material, the third field need not be sent at all as the decoder can easily repeat a field from memory. As a result the same compressed film material can be output at 50 or 60 Hz as required.

1.15 Some guidelines

Although compression techniques themselves are complex, there are some simple rules which can be used to avoid disappointment. Used wisely, compression has a number of advantages. Used in an inappropriate manner, disappointment is almost inevitable and the technology could get a bad name. The next few points are worth remembering.

- Compression technology may be exciting, but if it is not necessary it should not be used.
- If compression is to be used, the degree of compression should be as small as possible; i.e. use the highest practical bit rate.
- Cascaded compression systems cause loss of quality, and the lower the bit rates the worse this gets. Quality loss increases if any post production steps are performed between compressions.
- Compression systems cause delay.
- Compression systems work best with clean source material. Noisy signals or poorly decoded composite video give poor results.
- Compressed data are generally more prone to transmission errors than non-compressed data. The choice of a compression scheme must consider the error characteristics of the channel.
- Audio codecs need to be level calibrated so that when sound pressure level dependent decisions are made in the coder those levels actually exist at the microphone.
- Low-bit-rate coders should only be used for the final delivery of post-produced signals to the end user.
- Compression quality can only be assessed subjectively.
- Don't believe statements comparing codec performance to 'VHS quality' or similar. Compression artifacts are quite different from the artifacts of consumer VCRs.
- Quality varies wildly with source material. Beware of 'convincing' demonstrations which may use selected material to achieve low bit rates. Use your own test material, selected for a balance of difficulty.
- Don't be browbeaten by the technology. You don't have to understand it to assess the results. Your eyes and ears are as good as anyone's, so don't be afraid to criticize artifacts. In the case of video, use still frames to distinguish spatial artifacts from temporal artifacts.

Reference

1. HUFFMAN, D.A., A method for the construction of minimum redundancy codes. *Proc. IRE*, **40**, 1098–1101 (1952)

Chapter 2
Fundamentals

2.1 What is an audio signal?

Actual sounds are converted to electrical signals for convenience of handling, recording and conveying from one place to another. This is the job of the microphone. There are two basic types of microphone: those which measure the variations in air pressure due to sound, and those which measure the air velocity due to sound, although there are numerous practical types which are a combination of both.

The sound pressure or velocity varies with time and so does the output voltage of the microphone, in proportion. The output voltage of the microphone is thus an analog of the sound pressure or velocity.

As sound causes no overall air movement, the average velocity of all sounds is zero, which corresponds to silence. As a result the bi-directional air movement gives rise to bipolar signals from the microphone, where silence is in the centre of the voltage range, and instantaneously negative or positive voltages are possible. Clearly the average voltage of all audio signals is also zero, and so when level is measured, it is necessary to take the modulus of the voltage, which is the job of the rectifier in the level meter. When this is done, the greater the amplitude of the audio signal, the greater the modulus becomes, and so a higher level is displayed.

Whilst the nature of an audio signal is very simple, there are many applications of audio, each requiring different bandwidth and dynamic range.

2.2 What is a video signal?

The goal of television is to allow a moving picture to be seen at a remote place. The picture is a two-dimensional image, which changes as a function of time. This is a three-dimensional information source where the dimensions are distance across the screen, distance down the screen and time.

Whilst telescopes convey these three dimensions directly, this cannot be done with electrical signals or radio transmissions, which are restricted to a single parameter varying with time.

The solution in film and television is to convert the three-dimensional moving image into a series of still pictures, taken at the frame rate, and then, in television only, the two-dimensional images are scanned as a series of lines to produce a single voltage varying with time which can be recorded or transmitted. Europe, the Middle East and the former Soviet Union use the scanning standard of 625/50, whereas the USA and Japan use 525/59.94.

2.3 Types of video

Figure 2.1 shows some of the basic types of analog colour video. Each of these types can, of course, exist in a variety of line standards. Since practical colour cameras generally have three separate sensors, one for each primary colour, an *RGB* system will exist at some stage in the internal workings of the camera, even if it does not emerge in that form. *RGB* consists of three parallel signals each having the same spectrum, and is used where the highest accuracy is needed, often for production of still pictures. Examples of this are paint systems and in computer-aided design (CAD) displays. *RGB* is seldom used for real-time video recording: there is no standard *RGB* recording format for post-production or broadcast. As the red, green and blue signals directly represent part of the image, this approach is known as component video.

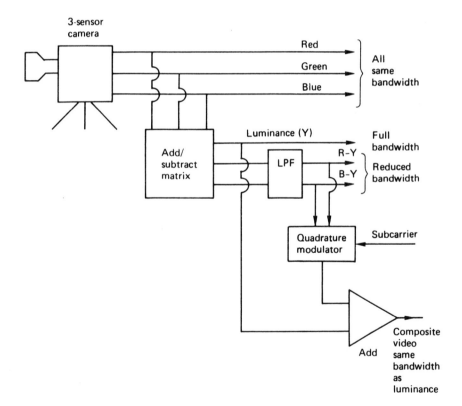

Figure 2.1 The major types of analog video. Red, green and blue signals emerge from the camera sensors, needing full bandwidth. If a luminance signal is obtained by a weighted sum of *R, G* and *B*, it will need full bandwidth, but the colour difference signals *R–Y* and *B–Y* need less bandwidth. Combining *R–Y* and *B–Y* into a subcarrier modulation scheme allows colour transmission in the same bandwidth as monochrome.

Some saving of bandwidth can be obtained by using colour difference working. The human eye relies on brightness to convey detail, and much less resolution is needed in the colour information. R, G and B are matrixed together to form a luminance (and monochrome compatible) signal Y which has full bandwidth. The matrix also produces two colour difference signals, $R-Y$ and $B-Y$, but these do not need the same bandwidth as Y; one-half or one-quarter will suffice, depending on the application. Colour difference signals represent an early application of perceptive coding; a saving in bandwidth is obtained by expressing the signals according to the way the eye operates. Analog colour difference recorders such as Betacam and M II record these signals separately. The D-1 and D-5 formats record 525/60 or 625/50 colour difference signals digitally and Digital Betacam does so using compression. In casual parlance, colour difference formats are often called component formats to distinguish them from composite formats.

For colour television broadcast in a single channel, the PAL, SECAM and NTSC systems interleave into the spectrum of a monochrome signal a subcarrier which carries two colour difference signals of restricted bandwidth. As the bandwidth required for composite video is no greater than that of luminance, it can be regarded as a form of compression performed in the analog domain. The artifacts which composite video introduces, and the inflexibility in editing resulting from the need to respect colour framing, serve as a warning that compression is not without its penalties. The subcarrier is intended to be invisible on the screen of a monochrome television set. A subcarrier-based colour system is generally referred to as composite video, and the modulated subcarrier is called chroma.

It is just not possible to compress composite video using modern transform-based coders. The embedded subcarrier would simply be destroyed in the process. Composite video compression is restricted to differential coding systems. The only possibilities for transform-based compression are RGB or colour difference signals. As RGB requires excessive bandwidth it makes no sense to use it with compression and so in practice only colour difference signals, which can be bandwidth reduced by perceptive coding, are used. Where signals to be compressed originate in composite form, they must be decoded first. The decoding must be performed as accurately as possible, with particular attention being given to the quality of the Y/C separation. The chroma in composite signals is deliberately designed to invert from frame to frame in order to lessen its visibility. Unfortunately, any residual chroma in luminance will be interpreted by inter-field compression systems as temporal luminance changes which need to be reproduced. This eats up data which should be used to render the picture. Residual chroma also results in high horizontal and vertical spatial frequencies in each field which appear to be wanted detail to the compressor.

2.4 What is a digital signal?

One of the vital concepts to grasp is that digital audio and video are simply alternative means of carrying the same information as their analog counterparts. An ideal digital recorder has the same characteristics as an ideal analog recorder: both of them are totally transparent and reproduce the original applied waveform without error. One need only compare high-quality analog and digital equipment side by side with the same signals to realize how transparent modern equipment

Fundamentals 19

can be. Needless to say, in the real world ideal conditions seldom prevail, so analog and digital equipment both fall short of the ideal. Digital equipment simply falls short of the ideal to a smaller extent than does analog and at lower cost, or, if the designer chooses, can have the same performance as analog at much lower cost. Compression is one of the techniques used to lower the cost, but it has the potential to lower the quality as well.

Any analog signal source can be characterized by a given useful bandwidth and signal-to-noise ratio. Video signals have very wide bandwidth extending over several megahertz but require only 50 dB or so SNR whereas audio signals require only 20 kHz but need much better SNR.

Although there are a number of ways in which audio and video waveforms can be represented digitally, there is one system, known as Pulse Code Modulation (PCM) which is in virtually universal use. Figure 2.2 shows how PCM works. Instead of being continuous, the time axis is represented in a discrete, or stepwise, manner. The waveform is not carried by continuous representation, but by measurement at regular intervals. This process is called sampling and the frequency with which samples are taken is called the sampling rate or sampling frequency F_s. The sampling rate is generally fixed and is not necessarily a function of any frequency in the signal, although in component video it will be line-locked for convenience. If every effort is made to rid the sampling clock of jitter, or time instability, every sample will be made at an exactly even time step. Clearly if there is any subsequent timebase error, the instants at which samples arrive will be changed and the effect can be detected. If samples arrive at some destination with an irregular timebase, the effect can be eliminated by storing the samples temporarily in a memory and reading them out using a stable, locally generated clock. This process is called timebase correction and is employed by all properly engineered digital systems. It should be stressed that sampling is an

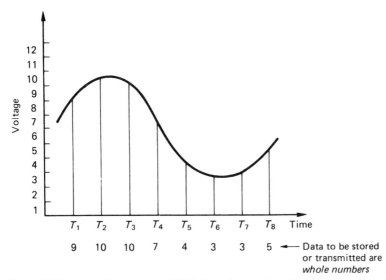

Figure 2.2 In pulse code modulation (PCM) the analog waveform is measured periodically at the sampling rate. The voltage (represented here by the height) of each sample is then described by a whole number. The whole numbers are stored or transmitted rather than the waveform itself.

analog process. Each sample still varies infinitely as the original waveform did.

Those who are not familiar with the technology often worry that sampling takes away something from a signal because it is not taking notice of what happened between the samples. This would be true in a system having infinite bandwidth, but no analog signal can have infinite bandwidth. All analog signal sources from microphones, tape decks, cameras and so on have a frequency response limit, as indeed do our ears and eyes. When a signal has finite bandwidth, the rate at which it can change is limited, and the way in which it changes becomes predictable. When a waveform can only change between samples in one way, it is then only necessary to carry the samples and the original waveform can be reconstructed from them.

Figure 2.2 also shows that each sample is also discrete, or represented in a stepwise manner. The length of the sample, which will be proportional to the voltage of the waveform, is represented by a whole number. This process is known as quantizing and results in an approximation, but the size of the error can be controlled until it is negligible. If, for example, we were to measure the height of humans to the nearest metre, virtually all adults would register two metres high and obvious difficulties would result. These are generally overcome by measuring height to the nearest centimetre. Clearly there is no advantage in going further and expressing our height in a whole number of millimetres or even micrometres. An appropriate resolution can be found just as readily for audio or video, and greater accuracy is not beneficial. The link between quality and sample resolution is explored later in this chapter. The advantage of using whole numbers is that they are not prone to drift. If a whole number can be carried from one place to another without numerical error, it has not changed at all. By describing waveforms numerically, the original information has been expressed in a way which is better able to resist unwanted changes.

Essentially, digital systems carry the original waveform numerically. The number of the sample is an analog of time, and the magnitude of the sample is an analog of the signal voltage. As both axes of the waveform are discrete, the waveform can be accurately restored from numbers as if it were being drawn on graph paper. If we require greater accuracy, we simply choose paper with smaller squares. Clearly more numbers are required and each one could change over a larger range.

Discrete numbers are used to represent the value of samples so that they can readily be transmitted or processed by binary logic. There are two ways in which binary signals can be used to carry sample data. When each digit of the binary number is carried on a separate wire this is called parallel transmission. The state of the wires changes at the sampling rate. This approach is used in the parallel video interfaces, as video needs a relatively short wordlength: eight or ten bits. Using multiple wires is cumbersome where a long wordlength is in use, and a single wire can be used where successive digits from each sample are sent serially. This is the definition of Pulse Code Modulation. Clearly the clock frequency must now be higher than the sampling rate.

If a well-engineered PCM digital channel having a wider bandwidth and a greater signal-to-noise ratio is put in series with an analog source, it is only necessary to set the levels correctly and the analog signal is then subject to no loss of information whatsoever. The digital clipping level is above the largest analog signal, the digital noise floor is below the inherent noise in the signal and

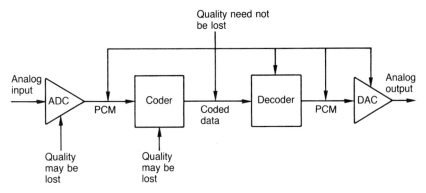

Figure 2.3 A typical digital compression system. The ADC and coder are responsible for most of the quality loss, whereas the PCM and coded data channels cause no further loss (excepting bit errors).

the low- and high-frequency response of the digital channel extends beyond the frequencies in the analog signal.

Digital signals of this form will be used as the input to compression systems and must also be output by the decoding stage in order that the signal can be returned to analog form. Figure 2.3 shows the stages involved. Between the coder and the decoder the signal is not PCM but will be in a format which is highly dependent on the kind of compression technique used. It will also be evident from Figure 2.3 where the signal quality of the system can be impaired. The PCM digital interfaces between the ADC and the coder and between the decoder and the DAC cause no loss of quality. Quality is determined by the ADC and by the performance of the coder. Generally, decoders do not cause loss of quality: they make the best of the data from the coder. Similarly, DACs cause little quality loss above that due to the ADC. In practical systems the loss of quality is dominated by the action of the coder. In communication theory, compression is known as *source coding* in order to distinguish it from the *channel coding* necessary to send data reliably down transmission or recording channels. This book is not concerned with channel coding, but details can be found elsewhere.[1]

2.5 Introduction to conversion

There are a number of ways in which an audio waveform can be digitally represented, but the most useful and therefore common is PCM as described above. The input is a continuous-time, continuous-voltage waveform, and this is converted into a discrete-time, discrete-voltage format by a combination of sampling and quantizing. As these two processes are orthogonal (a 64 dollar word meaning at right angles to one another) they are totally independent and can be performed in either order. Figure 2.4(a) shows an analog sampler preceding a quantizer, whereas Figure 2.4(b) shows an asynchronous quantizer preceding a digital sampler. Ideally, both will give the same results; in practice each suffers from different deficiencies. Both approaches will be found in real equipment.

The independence of sampling and quantizing allows each to be discussed quite separately in some detail, prior to combining the processes for a full understanding of conversion.

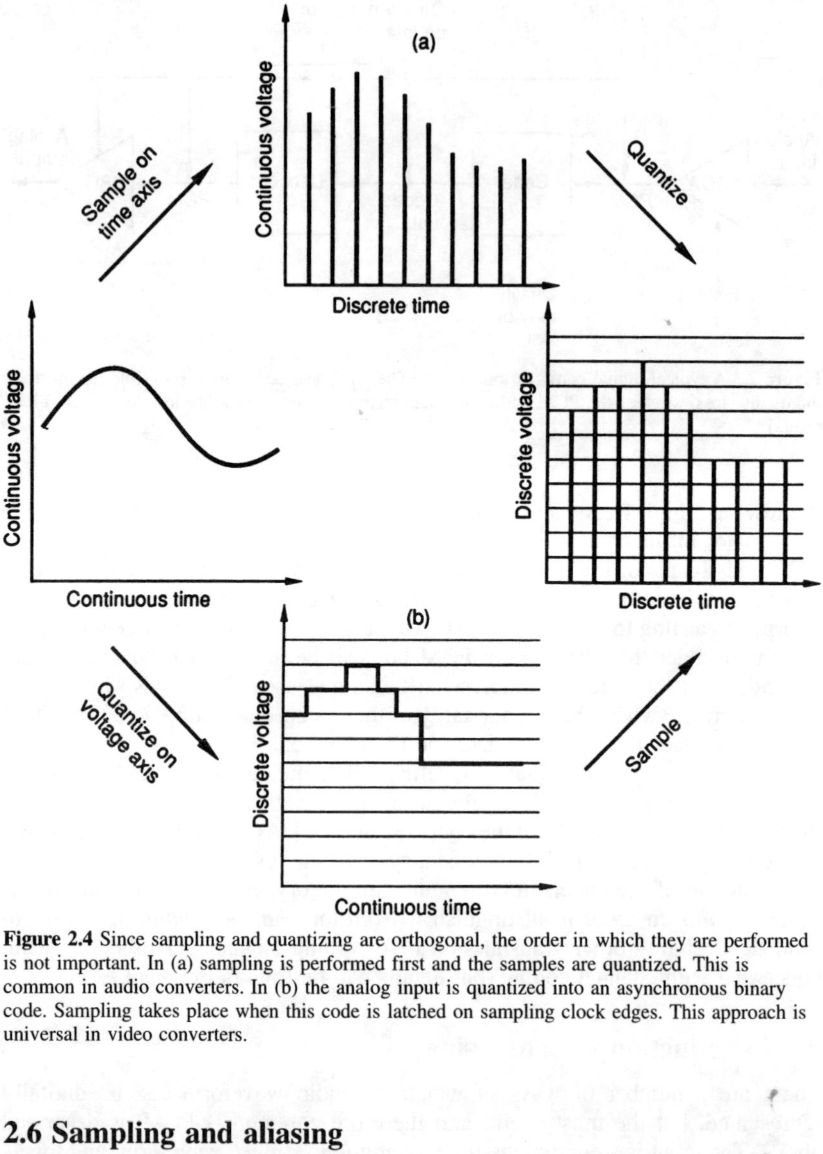

Figure 2.4 Since sampling and quantizing are orthogonal, the order in which they are performed is not important. In (a) sampling is performed first and the samples are quantized. This is common in audio converters. In (b) the analog input is quantized into an asynchronous binary code. Sampling takes place when this code is latched on sampling clock edges. This approach is universal in video converters.

2.6 Sampling and aliasing

Sampling is no more than periodic measurement, and it will be shown here that there is no theoretical need for sampling to be audible or visible. Practical equipment may, of course, be less than ideal, but, given good engineering practice, the ideal may be approached quite closely.

Sampling must be precisely regular, because the subsequent process of timebase correction assumes a regular original process. The sampling process originates with a pulse train which is shown in Figure 2.5(a) to be of constant amplitude and period. The signal waveform amplitude modulates the pulse train in much the same way as the carrier is modulated in an AM radio transmitter. One must be careful to avoid over-modulating the pulse train as shown in Figure

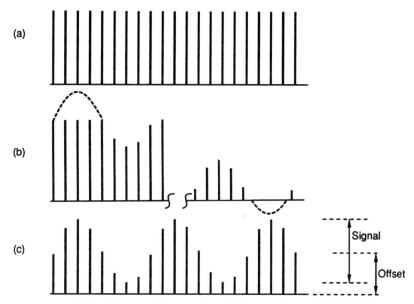

Figure 2.5 The sampling process requires a constant-amplitude pulse train as shown in (a). This is amplitude modulated by the waveform to be sampled. If the input waveform has excessive amplitude or incorrect level, the pulse train clips as shown in (b). For a bipolar waveform, the greatest signal level is possible when an offset of half the pulse amplitude is used to centre the waveform as shown in (c).

2.5(b), and in the case of bipolar signals this is helped by applying a DC offset to the analog waveform so that silence corresponds to a level half-way up the pulses as in Figure 2.5(c). Clipping due to any excessive input level will then be symmetrical.

In the same way that AM radio produces sidebands or images above and below the carrier, sampling also produces sidebands, although the carrier is now a pulse train and has an infinite series of harmonics as can be seen in Figure 2.6(a). The sidebands in Figure 2.6(b) repeat above and below each harmonic of the sampling rate.

The sampled signal can be returned to the continuous-time domain simply by passing it into a low-pass filter. This filter has a frequency response which prevents the images from passing, and only the baseband signal emerges, completely unchanged.

If an input is supplied having an excessive bandwidth for the sampling rate in use, the sidebands will overlap, and the result is aliasing, where certain output frequencies are not the same as their input frequencies but become difference frequencies. It will be seen from Figure 2.6(c) that aliasing occurs when the input frequency exceeds half the sampling rate, and this derives the most fundamental rule of sampling, which is that the sampling rate must be at least twice the highest input frequency.

In addition to the low-pass filter needed at the output to return to the continuous-time domain, a further low-pass filter is needed at the input to prevent aliasing. If input frequencies of more than half the sampling rate cannot reach the sampler, aliasing cannot occur.

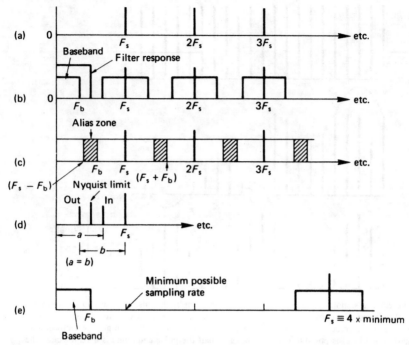

Figure 2.6 (a) Spectrum of sampling pulses. (b) Spectrum of samples. (c) Aliasing due to sideband overlap. (d) Beat-frequency production. (e) 4× oversampling.

Whilst aliasing has been described above in the frequency domain, it can equally be described in the time domain. In Figure 2.7(a) the sampling rate is obviously adequate to describe the waveform, but in Figure 2.7(b) it is inadequate and aliasing has occurred.

Aliasing is commonly seen on television and in the cinema, owing to the relatively low frame rates used. With a frame rate of 24 Hz, a film camera will alias on any object changing at more than 12 Hz. Such objects include the spokes of stagecoach wheels, especially when being chased by Indians! When the spoke-passing frequency reaches 24 Hz the wheels appear to stop.

In television systems the input image which falls on the camera sensor will be continuous in time, and continuous in two spatial dimensions corresponding to

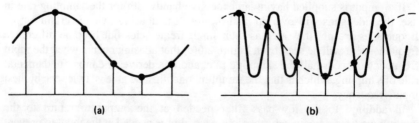

Figure 2.7 In (a), the sampling is adequate to reconstruct the original signal. In (b) the sampling rate is inadequate, and reconstruction produces the wrong waveform (detailed). Aliasing has taken place.

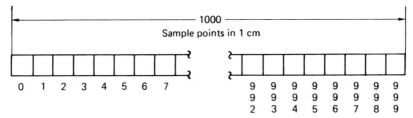

Figure 2.8 If the above spatial sampling arrangement of 1000 points per centimetre is scanned in 1 millisecond, the sampling rate will become 1 megahertz.

the height and width of the sensor. All three of these continuous dimensions will be sampled. There is a direct connection between the concept of temporal sampling, where the input signal changes with respect to time at some frequency and is sampled at some other frequency, and spatial sampling, where an image changes a given number of times per unit distance and is sampled at some other number of times per unit distance. The connection between the two is the process of scanning. Temporal frequency can be obtained by multiplying spatial frequency by the speed of the scan. Figure 2.8 shows a hypothetical image sensor which has 1000 discrete sensors across a width of 1 cm. The spatial sampling rate of this sensor is thus 1000 per centimetre. If the sensors are measured sequentially during a scan which takes one millisecond to go across the one centimetre width, the result will be a temporal sampling rate of 1 MHz.

The majority of today's television standards use 2:1 interlace. Figure 2.9(a) shows that in such a system there is an odd number of lines in a frame, and the frame is split into two fields. The first field begins with a whole line and ends with a half line, and the second field begins with a half line, which allows it to interleave spatially with the first field. The field rate is intended to determine the

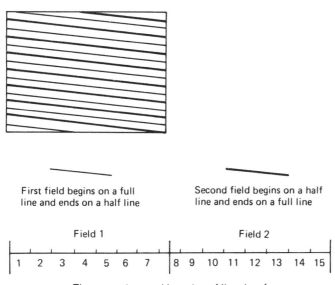

Figure 2.9(a) 2:1 interlace.

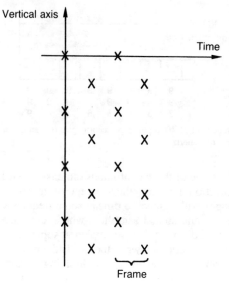

Figure 2.9(b) Vertical/temporal view of interlaced system.

flicker frequency, whereas the frame rate determines the bandwidth needed, which is thus halved along with the information rate. Information theory tells us that halving the information rate must reduce quality, and so the saving in bandwidth is accompanied by a variety of effects. Figure 2.9(b) shows the spatial/temporal sampling points in a 2:1 interlaced system. Figure 2.10 shows

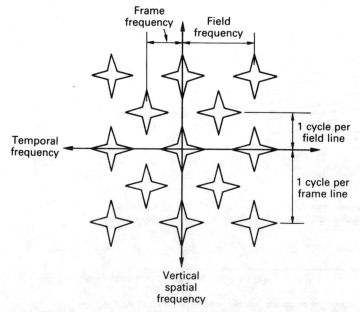

Figure 2.10 The vertical/temporal spectrum of monochrome video due to interlace.

the corresponding vertical/temporal spectrum which has a similar appearance. The anti-aliasing and reconstruction filters ideally should have a triangular passband. In practice these are seldom implemented. The triangular baseband reveals the confidence trick which is interlace: high spatial frequencies can be resolved in the absence of motion and rapid motion can be resolved in the absence of high spatial frequencies. Detail cannot be conveyed with motion; in practical systems lacking filters the result is aliasing. The poor motion portrayal of interlaced systems reflects in difficulties in performing accurate motion-compensated compression. The main drawback of interlaced systems is the triangular baseband spectrum. All filters required for processes such as standards conversion, resolution scaling or display upsampling must be two-dimensional; two one-dimensional filters cascaded can only offer a rectangular passband. The computer graphics community has been steadfastly ignoring interlace for a long time and with the steady merging of computer and video technology it may not be adopted in future television formats.

2.7 Reconstruction

If ideal low-pass anti-aliasing and anti-image filters are assumed, having a vertical cut-off slope at half the sampling rate, an ideal spectrum shown in Figure 2.11(a) is obtained. Figure 2.11(b) shows that the impulse response of a phase linear ideal low-pass filter is a $\sin x/x$ waveform in the time domain. Such a waveform passes through zero volts periodically. If the cut-off frequency of the filter is one-half of the sampling rate, the impulse passes through zero *at the sites of all other samples*. Thus at the output of such a filter, the voltage at the centre of a sample is due to that sample alone, since the value of *all* other samples is zero at that instant. In other words the continuous time output waveform must join up the tops of the input samples. In between the sample instants, the output of the filter is the sum of the contributions from many impulses, and the waveform smoothly joins the tops of the samples. If the time domain is being considered, the anti-image filter of the frequency domain can equally well be called the reconstruction filter. It is a consequence of the band-limiting of the original anti-aliasing filter that the filtered analog waveform could only travel between the sample points in one way. As the reconstruction filter has the same frequency response, the reconstructed output waveform must be identical to the original band-limited waveform prior to sampling. A rigorous mathematical proof of the above has been available since the 1930s, when PCM was invented, and can also be found in Betts.[2]

2.8 Filter design

The ideal filter with a vertical 'brick-wall' cut-off slope is difficult to implement. As the slope tends to vertical, the delay caused by the filter goes to infinity: the quality is marvellous but you don't live to measure it. In practice, a filter with a finite slope is accepted as shown in Figure 2.12, and the sampling rate has to be raised a little to prevent aliasing. There is no absolute factor by which the sampling rate must be raised; it depends upon the filters which are available.

It is not easy to specify such filters, particularly the amount of stopband rejection needed. The amount of aliasing resulting would depend on, among other things, the amount of out-of-band energy in the input signal. This is seldom

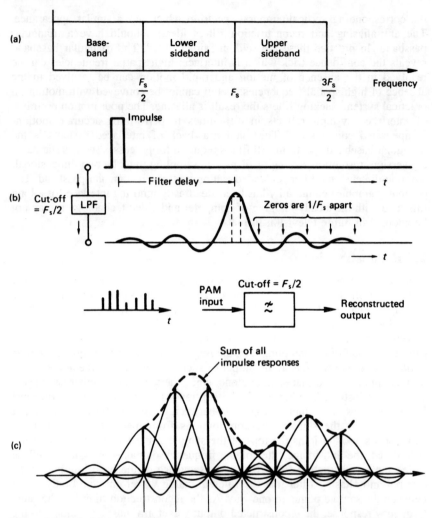

Figure 2.11 If ideal 'brick wall' filters are assumed, the efficient spectrum of (a) results. An ideal low-pass filter has an impulse response shown in (b). The impulse passes through zero at intervals equal to the sampling period. When convolved with a pulse train at the sampling rate, as shown in (c), the voltage at each sample instant is due to that sample alone as the impulses from all other samples pass through zero there.

a problem in video, but can warrant attention in audio where overspecified bandwidths are sometimes found. As a further complication, an out-of-band signal will be attenuated by the response of the anti-aliasing filter to that frequency, but the residual signal will then alias, and the reconstruction filter will reject it according to its attenuation at the new frequency to which it has aliased.

It could be argued that the reconstruction filter is unnecessary in audio, since all of the images are outside the range of human hearing. However, the slightest non-linearity in subsequent stages would result in gross intermodulation

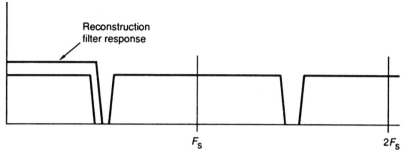

Figure 2.12 As filters with finite slope are needed in practical systems, the sampling rate is raised slightly beyond twice the highest frequency in the basband.

distortion. The possibility of damage to tweeters and beating with the bias systems of analog tape recorders must also be considered. In video the filters are essential to constrain the bandwidth of the signal to the allowable broadcast channel width.

Figure 2.13 shows the terminology used to describe the common elliptic low-pass filter. These filters are popular because they can be realized with fewer components than other filters of similar response. It is a characteristic of these elliptic filters that there are ripples in the passband and stopband. In much equipment the anti-aliasing filter and the reconstruction filter will have the same specification, so that the passband ripple is doubled with a corresponding increase in dispersion. Sometimes slightly different filters are used to reduce the effect.

It is difficult to produce an analog filter with low distortion. Passive filters using inductors suffer non-linearity at high levels due to the *B/H* curve of the cores. It seems a shame to go to such great lengths to remove the non-linearity of magnetic tape from a recording using digital techniques only to pass the signal through magnetic inductors in the filters. Active filters can simulate inductors which are linear using op-amp techniques, but they tend to suffer non-linearity at

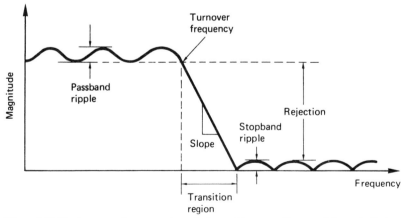

Figure 2.13 The important features and terminology of low-pass filters used for anti-aliasing and reconstruction.

high frequencies where the falling open-loop gain reduces the effect of feedback. Active filters can also contribute noise, but this is not necessarily a bad thing in controlled amounts, since it can act as a dither source (see Section 2.15).

It is instructive to examine the phase response of such filters. Since a sharp cut-off is generally achieved by cascading many filter sections which cut at a similar frequency; the phase responses of these sections will accumulate. The phase may start to leave linearity at a fraction of the cut-off frequency and, by the time this is reached, the phase may have completed several revolutions. Meyer[3] suggests that these phase errors are audible and that equalization is necessary. In video, phase linearity is essential as otherwise the different frequency components of a contrast step are smeared across the screen. An advantage of linear phase filters is that ringing is minimized, and there is less possibility of clipping on transients.

It is possible with substantial expense and effort to construct a ripple-free phase-linear filter with the required stopband rejection[4,5] but the requirement can be eliminated by the use of oversampling. As shown in Figure 2.14 a high sampling rate produces a large spectral gap between the baseband and the first lower sideband. The anti-aliasing and reconstruction filters need only have a gentle roll-off, causing minimum disturbance to phase linearity in the baseband, and the Butterworth configuration, which does not have ripple or dispersion, can be used. The penalty of oversampling is that an excessive data rate results. It is necessary to reduce the rate following the ADC using a digital LPF. The sampling rate can subsequently be raised prior to the DAC using an interpolator. Digital

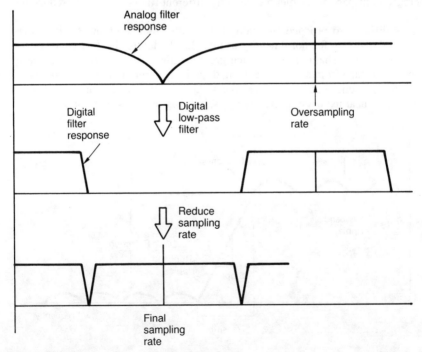

Figure 2.14 With oversampling the analog filters become non-critical since the passband is determined by digital filters.

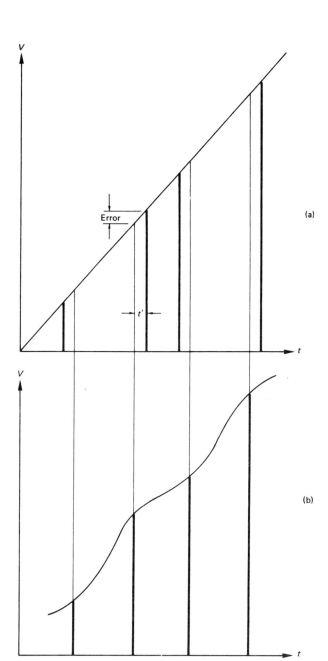

Figure 2.15 The effect of sampling timing jitter on noise, and calculation of the required accuracy for a 16 bit system. (a) Ramp sampled with jitter has error proportional to slope. (b) When jitter is removed by later circuits, error appears as noise added to samples. For a 16 bit system there are $2^{16}Q$, and the maximum slope at 20 kHz will be $20\,000\pi \times 2^{16}Q$ per second. If jitter is to be neglected, the noise must be less than $\tfrac{1}{2}Q$, thus timing accuracy t' multiplied by maximum slope $= \tfrac{1}{2}Q$ or $20\,000\pi \times 2^{16}Qt' = \tfrac{1}{2}Q$

$$\therefore t' = \frac{1}{2 \times 20\,000 \times \pi \times 2^{16}} = 121 \text{ ps}$$

filters can be made perfectly phase linear and, using LSI, can be inexpensive to construct. The superiority of oversampling converters means that they have become universal in audio and appear set to do so in video.

2.9 Sampling clock jitter

The instants at which samples are taken in an ADC and the instants at which DACs make conversions must be evenly spaced, otherwise unwanted signals can be added to the waveform. Figure 2.15(a) shows the effect of sampling clock jitter on a sloping waveform. Samples are taken at the wrong times. When these samples have passed through a system, the timebase correction stage prior to the DAC will remove the jitter, and the result is shown in Figure 2.15(b). The magnitude of the unwanted signal is proportional to the slope of the signal waveform and so increases with frequency. The nature of the unwanted signal depends on the spectrum of the jitter. If the jitter is random, the effect is noise-like and relatively benign unless the amplitude is excessive. Clock jitter is, however, not necessarily random. Figure 2.16 shows that one source of clock jitter is crosstalk on the clock signal. The unwanted additional signal changes the time at which the sloping clock signal appears to cross the threshold voltage of the clock receiver. The threshold itself may be changed by ripple on the clock receiver power supply. There is no reason why these effects should be random; they may be periodic and potentially discernible.

The allowable jitter is measured in picoseconds in both audio and video signals, as shown in Figure 2.15, and clearly steps must be taken to eliminate it by design. Converter clocks must be generated from clean power supplies which are well decoupled from the power used by the logic because a converter clock must have a good signal-to-noise ratio. If an external clock is used, it cannot be

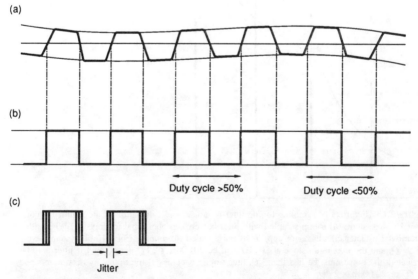

Figure 2.16 Crosstalk in transmission can result in unwanted signals being added to the clock waveform. It can be seen here that a low-frequency interference signal affects the slicing of the clock and causes a periodic jitter.

used directly, but must be fed through a well-damped phase-locked loop which will filter out the jitter. The external clock signal is sometimes fed into the clean circuitry using an optical coupler to improve isolation.

Although it has been documented for many years, attention to control of clock jitter is not as great in actual audio hardware as it might be. It accounts for much of the slight audible differences between converters reproducing the same data. A well-engineered converter should substantially reject jitter on an external clock and should sound the same when reproducing the same data irrespective of the source of the data. Jitter tends to be less noticeable on digital video signals and is generally not an issue until it becomes great enough to cause data errors.

2.10 Choice of audio sampling rate

The Nyquist criterion is only the beginning of the process which must be followed to arrive at a suitable sampling rate. The slope of available filters will compel designers to raise the sampling rate above the theoretical Nyquist rate. For consumer products, the lower the sampling rate the better, since the cost of the medium is directly proportional to the sampling rate: thus sampling rates near to twice 20 kHz are to be expected. For professional products, there is a need to operate at variable speed for pitch correction. When the speed of a digital recorder is reduced, the offtape sampling rate falls, and Figure 2.17 shows that with a minimal sampling rate the first image frequency can become low enough to pass the reconstruction filter. If the sampling frequency is raised without changing the response of the filters, the speed can be reduced without this problem. It follows that variable-speed recorders, generally those with stationary heads, must use a higher sampling rate.

In the early days of digital audio, video recorders were adapted to store audio samples by creating a pseudo-video waveform which could convey binary as black and white levels.[6] The sampling rate of such a system is constrained to relate simply to the field rate and field structure of the television standard used, so that an integer number of samples can be stored on each usable TV line in the field. Such a recording can be made on a monochrome recorder, and these recordings are made in two standards, 525 lines at 60 Hz and 625 lines at 50 Hz. Thus it was necessary to find a frequency which is a common multiple of the two and also suitable for use as a sampling rate.

The allowable sampling rates in a pseudo-video system can be deduced by multiplying the field rate by the number of active lines in a field (blanked lines cannot be used) and again by the number of samples in a line. By careful choice of parameters it is possible to use either 525/60 or 625/50 video with a sampling rate of 44.1 kHz.

In 60 Hz video, there are 35 blanked lines, leaving 490 lines per frame, or 245 lines per field for samples. If three samples are stored per line, the sampling rate becomes $60 \times 245 \times 3 = 44.1$ kHz. In 50 Hz video, there are 37 lines of blanking, leaving 588 active lines per frame, or 294 per field, so the same sampling rate is given by $50 \times 294 \times 3 = 44.1$ kHz. The sampling rate of 44.1 kHz came to be that of the Compact Disc. Even though CD has no video circuitry, the equipment used to make CD masters is video based and determines the sampling rate.

For landlines to FM stereo broadcast transmitters having a 15 kHz audio bandwidth, the sampling rate of 32 kHz is more than adequate, and has been in

34 Fundamentals

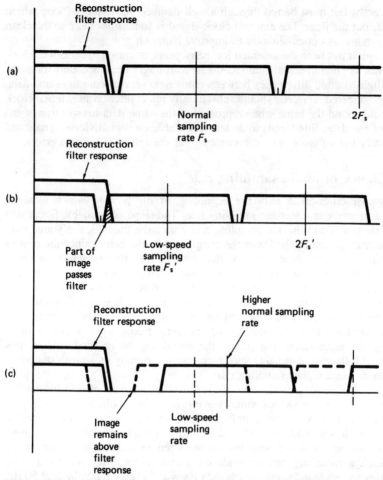

Figure 2.17 At normal speed, the reconstruction filter correctly prevents images entering the baseband, as in (a). When speed is reduced, the sampling rate falls, and a fixed filter will allow part of the lower sideband of the sampling frequency to pass. If the sampling rate of the machine is raised, but the filter characteristic remains the same, the problem can be avoided, as in (c).

use for some time in the United Kingdom and Japan. This frequency is also in use in the NICAM 728 stereo TV sound system. The professional sampling rate of 48 kHz was proposed as having a simple relationship to 32 kHz, being far enough above 40 kHz for variable-speed operation, and having a simple relationship with PAL video timing which would allow digital video recorders to store the convenient number of 960 audio samples per video field. This is the sampling rate used by all production DVTRs. The field rate offset of NTSC does not easily relate to any of the above sampling rates, and requires special handling which is outside the scope of this book.[7]

Although in a perfect world the adoption of a single sampling rate might have had virtues, for practical and economic reasons digital audio now has essentially three rates to support: 32 kHz for broadcast, 44.1 kHz for CD, and 48 kHz for professional use.[8] Variations of the DAT format will support all of these rates, although the 32 kHz version is uncommon. In MPEG-2 these audio sampling rates may be halved for low-bit-rate applications with a corresponding loss of audio bandwidth.

2.11 Video sampling structures

Component or colour difference signals are used primarily for post-production work where quality and flexibility are paramount. In colour difference working, the important requirement is for image manipulation in the digital domain. This is facilitated by a sampling rate which is a multiple of line rate because then there is a whole number of samples in a line and samples are always in the same position along the line and can form neat columns. A practical difficulty is that the line period of the 525 and 625 systems is slightly different. The problem was overcome by the use of a sampling clock which is an integer multiple of both line rates.

CCIR 601 recommends the use of certain sampling rates which are based on integer multiples of the carefully chosen fundamental frequency of 3.375 MHz. This frequency is normalized to 1 in the document.

In order to sample 625/50 luminance signals without quality loss, the lowest multiple possible is 4 which represents a sampling rate of 13.5 MHz. This frequency line-locks to give 858 samples per line period in 525/59.94 and 864 samples per line period in 625/50. The spectra of such sampled luminance signals are shown in Figure 2.18.

In the component analog domain, the colour difference signals used for production purposes typically have one-half the bandwidth of the luminance signal. Thus a sampling rate multiple of 2 is used and results in 6.75 MHz. This sampling rate allows respectively 429 and 432 samples per line.

Component video sampled in this way has a 4:2:2 format. Whilst other combinations are possible, 4:2:2 is the format for which the majority of digital component equipment is constructed and is the only component format for which parallel and serial interface standards exist. The D-1, D-5 and Digital Betacam DVTRs operate with 4:2:2 format data. Figure 2.19 shows the spatial

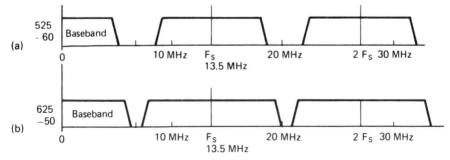

Figure 2.18 Spectra of video sampled at 13.5 MHz. In (a) the baseband 525/60 signal at left becomes the sidebands of the sampling rate and its harmonics. In (b) the same process for the 635/60 signal results in a smaller gap between baseband and sideband because of the wider bandwidth of the 625 system. The same sampling rate for both standards results in a great deal of commonality between 50 Hz and 60 Hz equipment.

36 Fundamentals

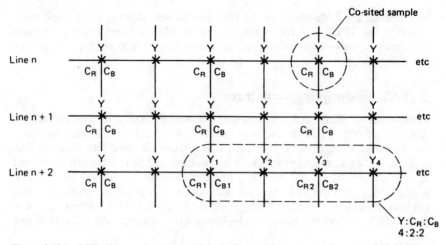

Figure 2.19 In CCIR-601 sampling mode 4:2:2, the line synchronous sampling rate of 13.5 MHz results in samples having the same position in successive lines, so that vertical columns are generated. The sampling rates of the colour difference signals C_R, C_B are one-half of that of luminance, i.e., 6.75 MHz, so that there are alternate Y only samples and co-sited samples which describe Y, C_R and C_B. In a run of four samples, there will be four Y samples, two C_R samples and two C_B samples, hence 4:2:2.

arrangement given by 4:2:2 sampling. Luminance samples appear at half the spacing of colour difference samples, and every other luminance sample is co-sited with a pair of colour difference samples. Co-siting is important because it allows all attributes of one picture point to be conveyed with a three-sample vector quantity. Modification of the three samples allows such techniques as colour correction to be performed. This would be difficult without co-sited information. Co-siting is achieved by clocking the three ADCs simultaneously. In some equipment one ADC is multiplexed between the two colour difference signals. In order to obtain co-sited data it will then be necessary to have an analog delay in one of the signals.

For lower bandwidths, particularly in pre-filtering operations prior to compression, multiples of 1 and 3 can also be used for colour difference and luminance respectively. 4:1:1 delivers colour bandwidth in excess of that required by the composite formats. 4:1:1 is used in the 525 line version of the DVC quarter-inch digital video format. 3:1:1 meets 525 line bandwidth requirements. The factors of 3 and 1 do not, however, offer a columnar structure and are inappropriate for quality post-production.

In 4:2:2 the colour difference signals are sampled horizontally at half the luminance sampling rate, yet the vertical colour difference sampling rates are the same as for luminance. Where bandwidth is important, it is possible to halve the vertical sampling rate of the colour difference signals as well. Figure 2.20 shows that in 4:2:0 sampling, the colour difference samples only exist on alternate lines so that the same vertical and horizontal resolution is obtained. 4:2:0 is used in the 625 line version of the DVC format.

There are two ways of handling 16:9 aspect ratio video. In the anamorphic approach both the camera and display scan wider but there is no change to the sampling rates employed and therefore the same data can be employed unchanged.

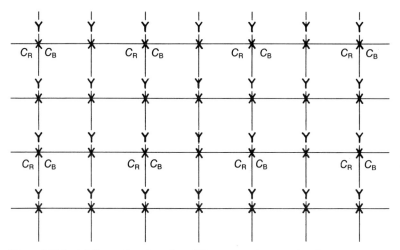

Figure 2.20 In 4:2:0 sampling the colour difference samples are only taken on alternate lines, making the vertical and horizontal resolution similar and halving the colour difference data rate.

Compared with 4:3, the horizontal spacing of the pixels in 16:9 must be greater as the same number are stretched across a wider picture. This results in a reduction of horizontal resolution, but standard 4:3 production equipment can be used subject to some modifications to the shape of pattern wipes in vision mixers. When viewed on a 4:3 monitor anamorphic signals appear squeezed horizontally.

In the second approach, the pixel spacing is kept the same as in 4:3 and the number of samples per active line must then be increased by 16:12. This requires the luminance sampling rate to rise to 18 MHz and the colour difference sampling rate becomes 9 MHz. The D-5 DVTR can record such signals. Composite video cannot be used for transform-based compression and so sampling rates for composite signals are not considered here.

2.12 The phase-locked loop

All digital video systems need to be clocked at the appropriate rate in order to function properly. Whilst a clock may be obtained from a fixed frequency oscillator such as a crystal, many operations in video require *genlocking* or synchronizing the clock to an external source.

In phase-locked loops, the oscillator can run at a range of frequencies according to the voltage applied to a control terminal. This is called a voltage-controlled oscillator or VCO. Figure 2.21 shows that the VCO is driven by a phase error measured between the output and some reference. The error changes the control voltage in such a way that the error is reduced, such that the output eventually has the same frequency as the reference. A low-pass filter is fitted in the control voltage path to prevent the loop becoming unstable. If a divider is placed between the VCO and the phase comparator, as in the figure, the VCO frequency can be made to be a multiple of the reference. This also has the effect of making the loop more heavily damped, so that it is less likely to change frequency if the input is irregular.

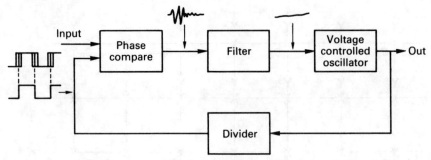

Figure 2.21 A phase-locked loop requires these components as a minimum. The filter in the control voltage serves to reduce clock jitter.

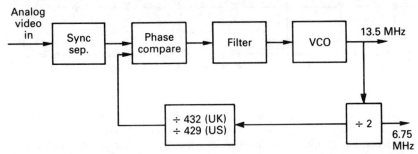

Figure 2.22 In order to obtain 13.5 MHz from input syncs, a PLL with an appropriate division ratio is required.

In digital video, the frequency multiplication of a phase-locked loop is extremely useful. Figure 2.22 shows how the 13.5 MHz clock of component digital video is obtained from the sync pulses of an analog reference by such a multiplication process.

2.13 Quantizing

Quantizing is the process of expressing some infinitely variable quantity by discrete or stepped values. Quantizing turns up in a remarkable number of everyday guises. Figure 2.23 shows that an inclined ramp enables infinitely

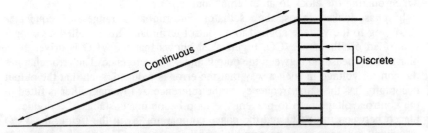

Figure 2.23 An analog parameter is continuous whereas a quantized parameter is restricted to certain values. Here the sloping side of a ramp can be used to obtain any height whereas a ladder only allows discrete heights.

variable height to be achieved, whereas a step-ladder allows only discrete heights to be had. A step-ladder quantizes height. When accountants round off sums of money to the nearest pound or dollar they are quantizing.

In audio the values to be quantized are infinitely variable voltages from an analog source. Strict quantizing is a process which is restricted to the voltage domain only. For the purpose of studying the quantizing of a single sample, time is assumed to stand still. This is achieved in practice either by the use of a track-hold circuit or the adoption of a quantizer technology which operates before the sampling stage.

Figure 2.24(a) shows that the process of quantizing divides the voltage range up into quantizing intervals Q. In applications such as telephony these may be of differing size, but for digital audio and video the quantizing intervals are made as identical as possible. If this is done, the binary numbers which result are truly proportional to the original analog voltage, and the digital equivalents of filtering and gain changing can be performed by adding and multiplying sample values. If the quantizing intervals are unequal this cannot be done. When all quantizing intervals are the same, the term uniform quantizing is used. The term linear quantizing will be found, but this is, like military intelligence, a contradiction in terms.

The term LSB (least significant bit) will also be found in place of quantizing interval in some treatments, but this is a poor term because quantizing is not always used to create binary values and because a bit can only have two values. In studying quantizing we wish to discuss values smaller than a quantizing interval, but a fraction of an LSB is a contradiction in terms.

Whatever the exact voltage of the input signal, the quantizer will determine the quantizing interval in which it lies. In what may be considered a separate step, the quantizing interval is then allocated a code value which is typically some form of binary number. The information sent is the number of the quantizing interval in which the input voltage lay. Exactly where that voltage lay within the interval is not conveyed, and this mechanism puts a limit on the accuracy of the quantizer. When the number of the quantizing interval is converted back to the analog domain, it will result in a voltage at the centre of the quantizing interval as this minimizes the magnitude of the error between input and output. The number range is limited by the wordlength of the binary numbers used. In a 16 bit system commonly used for audio, 65 536 different quantizing intervals exist, whereas video systems typically have 8 bit samples having 256 quantizing intervals.

2.14 Quantizing error

It is possible to draw a transfer function for such an ideal quantizer followed by an ideal DAC, and this is shown in Figure 2.24(b). A transfer function is simply a graph of the output with respect to the input. When the term linearity is used, this generally means the straightness of the transfer function. Linearity is a goal in audio and video, yet it will be seen that an ideal quantizer is anything but. Quantizing causes a voltage error in the sample which cannot exceed $\pm \frac{1}{2}Q$ unless the input is so large that clipping occurs.

Figure 2.24(b) shows that the transfer function is somewhat like a staircase, and the voltage corresponding to audio muting or video blanking is half-way up a quantizing interval, or on the centre of a tread. This is the so-called mid-tread quantizer which is universally used in audio and video. Figure 2.24(c) shows the

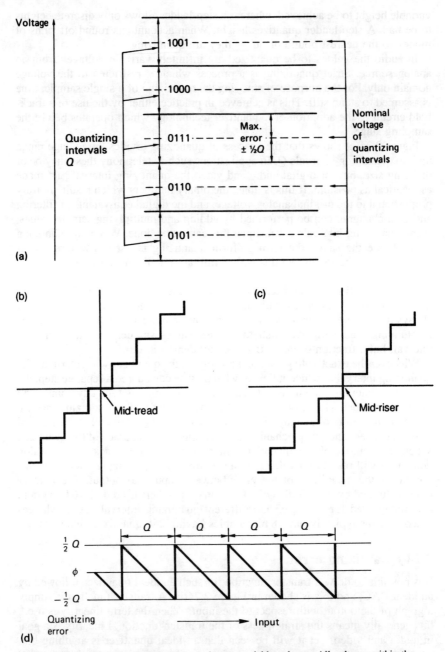

Figure 2.24 Quantizing assigns discrete numbers to variable voltages. All voltages within the same quantizing interval are assigned the same number which causes a DAC to produce the voltage at the centre of the intervals shown by the dashed lines in (a). This is the characteristic of the mid-tread quantizer shown in (b). An alternative system is the mid-riser system shown in (c). Here 0 volts analog falls between two codes and there is no code for zero. Such quantizing cannot be used prior to signal processing because the number is no longer proportional to the voltage. Quantizing error cannot exceed $\pm \frac{1}{2}Q$ as shown in (d).

alternative mid-riser transfer function which causes difficulty because it does not have a code value at muting/blanking level and as a result the code value is not proportional to the signal voltage.

In studying the transfer function it is better to avoid complicating matters with the aperture effect of the DAC. For this reason it is assumed here that output samples are of negligible duration. Then impulses from the DAC can be compared with the original analog waveform and the difference will be impulses representing the quantizing error waveform. As can be seen in Figure 2.25 the quantizing error waveform can be thought of as an unwanted signal which the

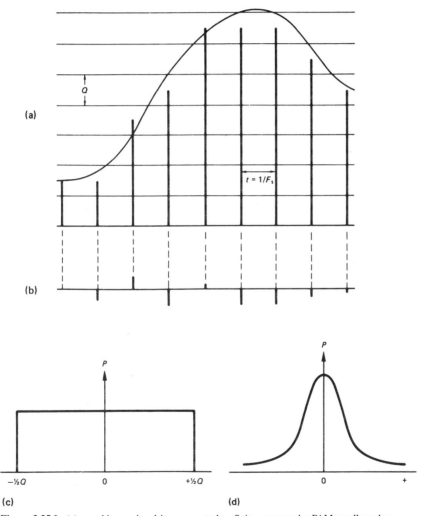

Figure 2.25 In (a) an arbitrary signal is represented to finite accuracy by PAM needles whose peaks are at the centre of the quantizing intervals. The errors caused can be thought of as an unwanted signal (b) added to the original. In (c) the amplitude of a quantizing error needle will be from $-\tfrac{1}{2}Q$ to $+\tfrac{1}{2}Q$ with equal probability. Note, however, that white noise in analog circuits generally has Gaussian amplitude distribution, shown in (d).

quantizing process adds to the perfect original. As the transfer function is non-linear, ideal quantizing can cause distortion. As a result practical digital audio devices use non-ideal quantizers to achieve linearity. The quantizing error of an ideal quantizer is a complex function, and it has been researched in great depth.[9]

As the magnitude of the quantizing error is limited, its effect can be minimized by making the signal larger. This will require more quantizing intervals and more bits to express them. The number of quantizing intervals multiplied by their size gives the quantizing range of the converter. A signal outside the range will be clipped. Clearly if clipping is avoided, the larger the signal the less will be the effect of the quantizing error.

Consider first the case where the input signal exercises the whole quantizing range and has a complex waveform. In audio this might be orchestral music; in video a bright, detailed contrasty scene. In these cases successive samples will have widely varying numerical values and the quantizing error on a given sample will be independent of that on others. In this case the size of the quantizing error will be distributed with equal probability between the limits. Figure 2.25(c) shows the resultant uniform probability density. In this case the unwanted signal added by quantizing is an additive broadband noise uncorrelated with the signal, and it is appropriate in this case to call it quantizing noise. This is not quite the same as thermal noise, which has a Gaussian probability shown in Figure 2.25(d). The subjective difference is slight. Treatments which then assume that quantizing error is *always* noise give results which are at variance with reality. Such approaches only work if the probability density of the quantizing error is uniform. Unfortunately at low levels, and particularly with pure or simple waveforms, this is simply not true.

At low levels, quantizing error ceases to be random, and becomes a function of the input waveform and the quantizing structure. Once an unwanted signal becomes a deterministic function of the wanted signal, it has to be classed as a distortion rather than a noise. We predicted a distortion because of the non-linearity or staircase nature of the transfer function. With a large signal, there are so many steps involved that we must stand well back, and a staircase with many steps appears to be a slope. With a small signal there are few steps and they can no longer be ignored.

The non-linearity of the transfer function results in distortion, which produces harmonics. Unfortunately these harmonics are generated *after* the anti-aliasing filter, and so any which exceed half the sampling rate will alias. Figure 2.26 shows how this results in anharmonic distortion in audio. These anharmonics result in spurious tones known as birdsinging. When the sampling rate is a multiple of the input frequency the result is harmonic distortion. Where more than one frequency is present in the input, intermodulation distortion occurs, which is known as granulation.

As the input signal is further reduced in level, it may remain within one quantizing interval. The output will be silent because the signal is now the quantizing error. In this condition, low-frequency signals such as air-conditioning rumble can shift the input in and out of a quantizing interval so that the quantizing distortion comes and goes, resulting in noise modulation.

In video, quantizing error in luminance results in visible contouring on low key scenes or flat fields. Slowly changing brightness across the screen is replaced by areas of constant brightness separated by sudden steps. In colour difference

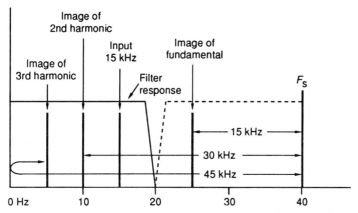

Figure 2.26 Quantizing produces distortion *after* the anti-aliasing filter; thus the distortion products will fold back to produce anharmonics in the audio band. Here the fundamental of 15 kHz produces second and third harmonic distortion at 30 and 45 kHz. This results in aliased products at 40 − 30 = 10 kHz and 40 − 45 = (−)5 kHz.

signals, contouring results in an effect known as posterization where subtle variations in colour are removed and large areas are rendered by the same colour as if they had been painted by numbers.

2.15 Dither

At high signal level, quantizing error is effectively noise. As the level falls, the quantizing error of an ideal quantizer becomes more strongly correlated with the signal and the result is distortion. If the quantizing error can be decorrelated from the input in some way, the system can remain linear. Dither performs the job of decorrelation by making the action of the quantizer unpredictable.

The first documented use of dither was in picture coding.[10] In this system, the noise added prior to quantizing was subtracted after reconversion to analog. This is known as subtractive dither. Although subsequent subtraction has some slight advantages,[9] it suffers from practical drawbacks, since the original noise waveform must accompany the samples or must be synchronously re-created at the DAC. This is virtually impossible in a system where the signal may have been edited. Practical systems use non-subtractive dither where the dither signal is added prior to quantization and no subsequent attempt is made to remove it. The introduction of dither inevitably causes a slight reduction in the signal-to-noise ratio attainable, but this reduction is a small price to pay for the elimination of non-linearities. As linearity is an essential requirement for digital audio and video, the use of dither is equally essential.

The ideal (noiseless) quantizer of Figure 2.25 has fixed quantizing intervals and must always produce the same quantizing error from the same signal. In Figure 2.27 it can be seen that an ideal quantizer can be dithered by linearly adding a controlled level of noise either to the input signal or to the reference voltage which is used to derive the quantizing intervals. There are several ways of considering how dither works, all of which are valid.

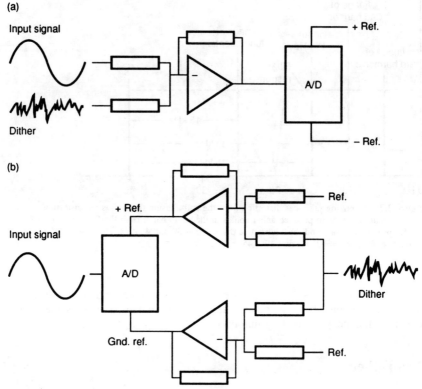

Figure 2.27 Dither can be applied to a quantizer in one of two ways. In (a) the dither is linearly added to the analog input signal, whereas in (b) it is added to the reference voltages of the quantizer.

The addition of dither means that successive samples effectively find the quantizing intervals in different places on the voltage scale. The quantizing error becomes a function of the dither, rather than just a function of the input signal. The quantizing error is not eliminated, but the subjectively unacceptable distortion is converted into broadband noise which is more benign.

An alternative way of looking at dither is to consider the situation where a low-level input signal is changing slowly within a quantizing interval. Without dither, the same numerical code results, and the variations within the interval are lost. Dither has the effect of forcing the quantizer to switch between two or more states. The higher the voltage of the input signal within the interval, the more probable it becomes that the output code will take on a higher value. The lower the input voltage within the interval, the more probable it is that the output code will take the lower value. The dither has resulted in a form of duty cycle modulation, and the resolution of the system has been extended indefinitely instead of being limited by the size of the steps.

Dither can also be understood by considering the effect it has on the transfer function of the quantizer. This is normally a perfect staircase, but in the presence of dither it is smeared horizontally until with a certain minimum amplitude the average transfer function becomes straight.

The characteristics of the noise used are rather important for optimal performance, although many sub-optimal but nevertheless effective systems are in use. The main parameters of interest are the peak-to-peak amplitude, and the probability distribution of the amplitude. Triangular probability works best and this can be obtained by summing the output of two uniform probability processes.

The use of dither invalidates the conventional calculations of signal-to-noise ratio available for a given wordlength. This is of little consequence as the rule of thumb that multiplying the number of bits in the wordlength by 6 dB gives the SNR gives a result that will be close enough for all practical purposes.

It has only been possible to introduce the principles of conversion of audio and video signals here. For more details of the operation of converters the reader is referred elsewhere.[1,11]

2.16 Binary codes for audio

For audio use, the prime purpose of binary numbers is to express the values of the samples which represent the original analog sound-pressure waveform. There will be a fixed number of bits in the sample, which determines the number range. In a 16 bit code there are 65 536 different numbers. Each number represents a different analog signal voltage, and care must be taken during conversion to ensure that the signal does not go outside the converter range, or it will be clipped. In Figure 2.28 it will be seen that in a simple system, the number range goes from 0000 hex, which represents the largest negative voltage, through 7FFF hex, which represents the smallest negative voltage, through 8000 hex, which

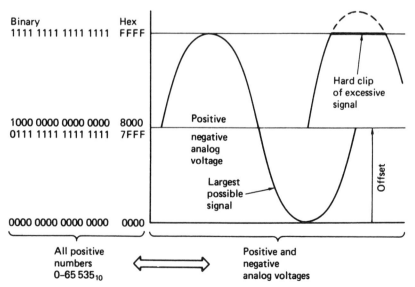

Figure 2.28 Offset binary coding is simple but causes problems in digital audio processing. It is seldom used.

46 Fundamentals

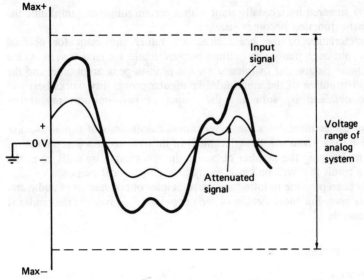

Figure 2.29 Attenuation of an audio signal takes place with respect to midrange.

represents the smallest positive voltage, to FFFF hex, which represents the largest positive voltage. Effectively, the number range of the converter has been shifted so that positive and negative voltages in a real audio signal can be expressed by binary numbers which are only positive. This approach is called offset binary, and is perfectly acceptable where the signal has been digitized only for recording or transmission from one place to another, after which it will be

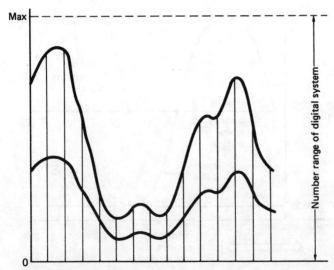

Figure 2.30 The result of an attempted attenuation in pure binary code is an offset. Pure binary cannot be used for digital audio processing.

converted back to analog. Under these conditions it is not necessary for the quantizing steps to be uniform, provided both ADC and DAC are constructed to the same standard. In practice, it is the requirements of signal processing in the digital domain which make both non-uniform quantizing and offset binary unsuitable.

Figure 2.29 shows that an audio signal voltage is referred to midrange. The level of the signal is measured by how far the waveform deviates from midrange, and attenuation, gain and mixing all take place around midrange. It is necessary to add sample values from two or more different sources to perform the mixing function, and adding circuits assume that all bits represent the same quantizing interval so that the sum of two sample values will represent the sum of the two original analog voltages. In non-uniform quantizing this is not the case, and such signals cannot readily be processed. If two offset binary sample streams are added together in an attempt to perform digital mixing, the result will be an offset which may lead to an overflow. Similarly, if an attempt is made to attenuate by, say, 6 dB by dividing all of the sample values by two, Figure 2.30 shows that a further offset results. The problem is that offset binary is referred to one end of

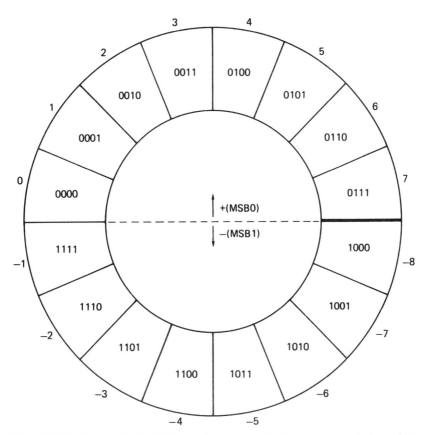

Figure 2.31 In this example of a 4 bit two's complement code, the number range is from −8 to +7. Note that the MSB determines polarity.

48 Fundamentals

the range. What is needed is a numbering system which operates symmetrically about the centre of the range.

In the two's complement system, the upper half of the pure binary number range has been redefined to represent negative quantities. If a pure binary counter is constantly incremented and allowed to overflow, it will produce all the numbers in the range permitted by the number of available bits, and these are shown for a 4 bit example drawn around the circle in Figure 2.31. As a circle has no real beginning, it is possible to consider it to start wherever it is convenient. In two's complement, the quantizing range represented by the circle of numbers does not start at zero, but starts on the diametrically opposite side of the circle. Zero is midrange, and all numbers with the MSB (most significant bit) set are considered negative. The MSB is thus the equivalent of a sign bit where 1 = minus. Two's complement notation differs from pure binary in that the most significant bit is inverted in order to achieve the half-circle rotation.

Figure 2.32 shows how a real ADC is configured to produce two's complement output. In Figure 2.32(a) an analog offset voltage equal to one-half the quantizing range is added to the bipolar analog signal in order to make it unipolar as at Figure 2.32(b). The ADC produces positive-only numbers at Figure 2.32(c) which are proportional to the input voltage. The MSB is then inverted at Figure 2.32(d) so that the all-zeros code moves to the centre of the quantizing range. The analog offset is often incorporated in the ADC as is the MSB inversion. Some converters are designed to be used in either pure binary or two's complement

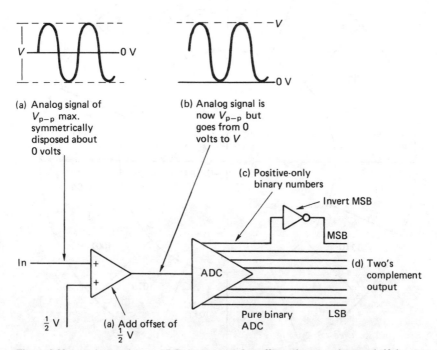

Figure 2.32 A two's complement ADC. At (a) an analog offset voltage equal to one-half the quantizing range is added to the bipolar analog signal in order to make it unipolar as at (b). The ADC produces positive-only numbers at (c), but the MSB is then inverted at (d) to give a two's complement output.

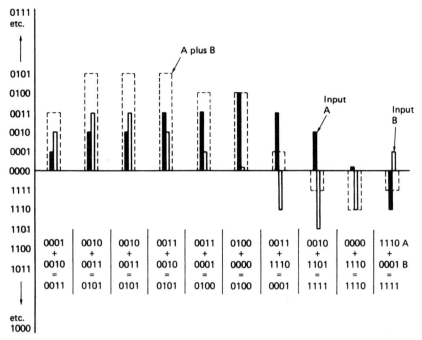

Figure 2.33 Using two's complement arithmetic, single values from two waveforms are added together with respect to midrange to give a correct mixing function.

mode. In this case the designer must arrange the appropriate DC conditions at the input. The MSB inversion may be selectable by an external logic level.

The two's complement system allows two sample values to be added, or mixed in audio parlance, and the result will be referred to the system midrange; this is analogous to adding analog signals in an operational amplifier. Figure 2.33 illustrates how adding two's complement samples simulates the audio mixing process. The waveform of input A is depicted by solid black samples, and that of B by samples with a solid outline. The result of mixing is the linear sum of the two waveforms obtained by adding pairs of sample values. The dashed lines depict the output values. Beneath each set of samples is the calculation which will be seen to give the correct result. Note that the calculations are pure binary. No special arithmetic is needed to handle two's complement numbers.

Figure 2.34 shows some audio waveforms at various levels with respect to the coding values. Where an audio waveform just fits into the quantizing range without clipping it has a level which is defined as 0 dBFs where Fs indicates *full scale*. Reducing the level by 6.02 dB makes the signal half as large and results in the second bit in the sample becoming the same as the sign bit. Reducing the level by a further 6.02 dB to −12 dBFs will make the second and third bits the same as the sign bit and so on. If a signal at −36 dBFs is input to a 16 bit system, only ten bits will be active; the remainder will copy the sign bit. For the best performance, analog inputs to digital systems must have sufficient levels to exercise the whole quantizing range.

50 Fundamentals

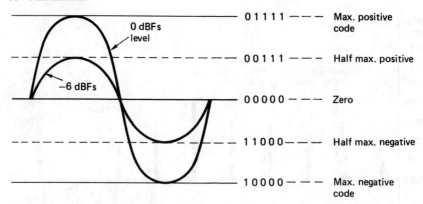

Figure 2.34 0 dBFs is defined as the level of the largest sinusoid which will fit into the quantizing range without clipping.

Using inversion, signal subtraction can be performed using only adding logic. The inverted input is added to perform a subtraction, just as in the analog domain. This permits a significant saving in hardware complexity, since only carry logic is necessary and no borrow mechanism need be supported.

In summary, two's complement notation is the most appropriate scheme for bipolar signals, and allows simple mixing in conventional binary adders. It is in virtually universal use in digital audio processing, and is accordingly adopted by all the major digital audio interfaces and recording formats. Two's complement format is also adopted for transform coding of both video and audio so that bipolar coefficients can be handled.

Two's complement numbers can have a radix point and bits below it just as pure binary numbers can. It should, however, be noted that in two's complement, if a radix point exists, numbers to the right of it are added. For example 1100.1 is not -4.5, it is $-4 + 0.5 = -3.5$.

2.17 Binary codes for component video

In video it is important to standardize the relationship between the absolute analog voltage of the waveform and the digital code value used to express it so that all machines will interpret the numerical data in the same way. These relationships are in the voltage domain and are independent of the scanning standard used.

Figure 2.35 shows how the luminance signal fits into the quantizing range of an 8 bit system. Numbering for 10 bit systems is shown with figures for 8 bits in brackets. Black is at a level of 64_{10} (16_{10}) and peak white is at 940_{10} (235_{10}) so that there is some tolerance of imperfect analog signals. The sync pulse will clearly go outside the quantizing range, but this is of no consequence as conventional syncs are not transmitted. The visible voltage range fills the quantizing range and this gives the best possible resolution.

The colour difference signals use offset binary, where 512_{10} (128_{10}) is the equivalent of blanking voltage. The peak analog limits are reached at 64_{10} (16_{10})

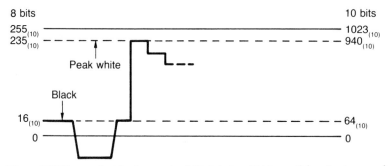

Figure 2.35 The standard luminance signal fits into 8 or 10 bit quantizing structures as shown here.

and 960_{10} (240_{10}) respectively, allowing once more some latitude for maladjusted analog inputs.

Note that the code values corresponding to the eight most significant bits being all ones or all zeros (i.e. the two extreme ends of the quantizing range in 8 bit data) are not allowed to occur in the active line as they are reserved for synchronizing. Converters must be followed by circuitry which catches these values and forces the LSB to a different value if out-of-range analog inputs are applied. The use of these reserved codes in digital interfaces precludes the use of two's complement coding, but, as was shown above, the offset is exactly half scale and so the data can be expressed in two's complement by inverting the MSB.

The peak-to-peak amplitude of Y is 880 (220) quantizing intervals, whereas for the colour difference signals it is 900 (225) intervals. There is thus a small gain difference between the signals. This will be cancelled out by the opposing gain difference at any future DAC, but must be borne in mind when digitally converting to other standards.

Although the coding standards of digital video interfaces use offset binary, for the purposes of transform coding the video must be converted to two's complement. The colour difference signals simply require the MSB to be inverted, whereas the luminance signal has an offset of half full scale subtracted first. This corresponds to 128 in 8 bit systems and 512 in 10 bit systems.

2.18 Introduction to digital processes

However complex a digital process, it can be broken down into smaller stages until finally one finds that there are really only two basic types of element in use, and these can be combined in some way and supplied with a clock to implement virtually any process. Figure 2.36 shows that the first type is a *logical* element. This produces an output which is a logical function of the input with minimal delay. The second type is a *storage* element which samples the state of the input(s) when clocked and holds or delays that state. The strength of binary logic is that the signal has only two states, and considerable noise and distortion of the binary waveform can be tolerated before the state becomes uncertain. At every logical element, the signal is compared with a threshold, and thus can pass

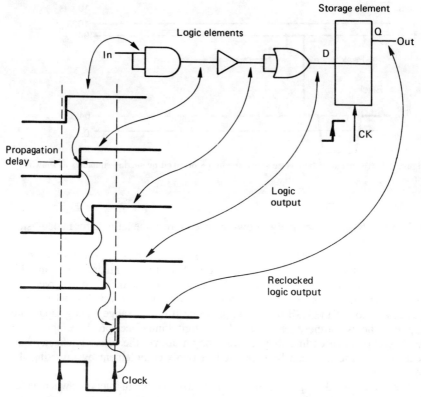

Figure 2.36 Logic elements have a finite propagation delay between input and output and cascading them delays the signal an arbitrary amount. Storage elements sample the input on a clock edge and can return a signal to near coincidence with the system clock. This is known as reclocking. Reclocking eliminates variations in propagation delay in logic elements.

through any number of stages without being degraded. In addition, the use of a storage element at regular locations throughout logic circuits eliminates time variations or jitter. Figure 2.36 shows that if the inputs to a logic element change, the output will not change until the *propagation delay* of the element has elapsed. However, if the output of the logic element forms the input to a storage element, the output of that element will not change until the input is sampled *at the next clock edge*. In this way the signal edge is aligned to the system clock and the propagation delay of the logic becomes irrelevant. The process is known as reclocking.

2.19 Logic elements

The two states of the signal when measured with an oscilloscope are simply two voltages, usually referred to as high and low. The actual voltage levels will depend on the type of logic family in use, and on the supply voltage used. Within logic, these levels are not of much consequence, and it is only necessary to know them when interfacing between different logic families or when driving external devices. The pure logic designer is not interested at all in these voltages, only in their meaning. Just as the electrical waveform from a microphone represents sound velocity, so the waveform in a logic circuit represents the truth of some statement. As there are only two states, there can only be *true* or *false* meanings. The true state of the signal can be assigned by the designer to either voltage state. When a high voltage represents a true logic condition and a low voltage represents a false condition, the system is known as *positive* logic, or *high true* logic. This is the usual system, but sometimes the low voltage represents the true condition and the high voltage represents the false condition. This is known as *negative* logic or *low true* logic. Provided that everyone is aware of the logic convention in use, both work equally well.

Positive logic name	Boolean expression	Positive logic symbol	Positive logic truth table	Plain English
Inverter or NOT gate	$Q = \bar{A}$	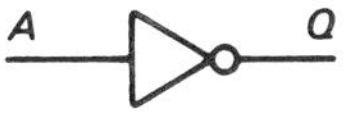	$\begin{array}{c\|c} A & Q \\ \hline 0 & 1 \\ 1 & 0 \end{array}$	Output is opposite of input
AND gate	$Q = A \cdot B$	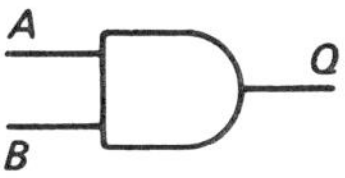	$\begin{array}{cc\|c} A & B & Q \\ \hline 0 & 0 & 0 \\ 0 & 1 & 0 \\ 1 & 0 & 0 \\ 1 & 1 & 1 \end{array}$	Output true when both inputs are true only
NAND (Not AND) gate	$Q = \overline{A \cdot B}$ $= \bar{A} + \bar{B}$	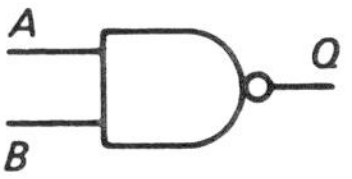	$\begin{array}{cc\|c} A & B & Q \\ \hline 0 & 0 & 1 \\ 0 & 1 & 1 \\ 1 & 0 & 1 \\ 1 & 1 & 0 \end{array}$	Output false when both inputs are true only
OR gate	$Q = A + B$	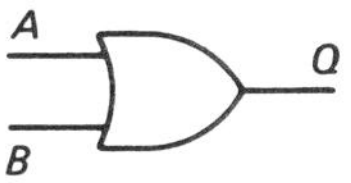	$\begin{array}{cc\|c} A & B & Q \\ \hline 0 & 0 & 0 \\ 0 & 1 & 1 \\ 1 & 0 & 1 \\ 1 & 1 & 1 \end{array}$	Output true if either or both inputs true
NOR (Not OR) gate	$Q = \overline{A + B}$ $= \bar{A} \cdot \bar{B}$	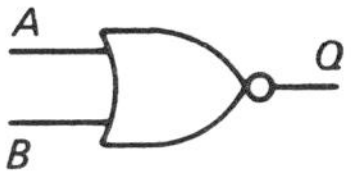	$\begin{array}{cc\|c} A & B & Q \\ \hline 0 & 0 & 1 \\ 0 & 1 & 0 \\ 1 & 0 & 0 \\ 1 & 1 & 0 \end{array}$	Output false if either or both inputs true
Exclusive OR (XOR) gate	$Q = A \oplus B$	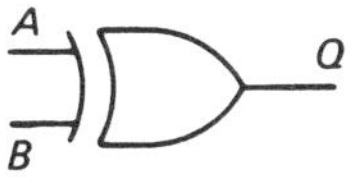	$\begin{array}{cc\|c} A & B & Q \\ \hline 0 & 0 & 0 \\ 0 & 1 & 1 \\ 1 & 0 & 1 \\ 1 & 1 & 0 \end{array}$	Output true if inputs are different

Figure 2.37 The basic logic gates compared.

In logic systems, all logical functions, however complex, can be configured from combinations of a few fundamental logic elements or *gates*. It is not profitable to spend too much time debating which are the truly fundamental ones, since most can be made from combinations of others. Figure 2.37 shows the important simple gates and their derivatives, and introduces the logical expressions to describe them, which can be compared with the truth-table notation. The figure also shows the important fact that when negative logic is used, the OR gate function interchanges with that of the AND gate.

If numerical quantities need to be conveyed down the two-state signal paths described here, then the only appropriate numbering system is binary, which has only two symbols, 0 and 1. Just as positive or negative logic could be used for the truth of a logical binary signal, it can also be used for a numerical binary signal. Normally, a high voltage level will represent a binary 1 and a low voltage will represent a binary 0, described as a 'high for a one' system. Clearly a 'low for a one' system is just as feasible. Decimal numbers have several columns, each of which represents a different power of ten; in binary the column position specifies the power of two.

2.20 Storage elements

The basic memory element in logic circuits is the latch, which is constructed from two gates as shown in Figure 2.38(a), and which can be set or reset. A more useful variant is the D-type latch shown in Figure 2.38(b) which remembers the state of the input at the time a separate clock either changes state for an edge-triggered device, or after it goes false for a level-triggered device. D-type latches are commonly available with four or eight latches to the chip. A shift register can be made from a series of latches by connecting the Q output of one latch to the D input of the next and connecting all of the clock inputs in parallel. Data are delayed by the number of stages in the register. Shift registers are also useful for converting between serial and parallel data transmissions.

Where large numbers of bits are to be stored, cross-coupled latches are less suitable because they are more complicated to fabricate inside integrated circuits than dynamic memory, and consume more current.

In large random access memories (RAMs), the data bits are stored as the presence or absence of charge in a tiny capacitor as shown in Figure 2.38(c). The capacitor is formed by a metal electrode, insulated by a layer of silicon dioxide from a semiconductor substrate, hence the term MOS (Metal Oxide Semiconductor). The charge will suffer leakage, and the value would become indeterminate after a few milliseconds. Where the delay needed is less than this, decay is of no consequence, as data will be read out before they have had a chance to decay. Where longer delays are necessary, such memories must be refreshed periodically by reading the bit value and writing it back to the same place. Most modern MOS RAM chips have suitable circuitry built in. Large RAMs store thousands of bits, and it is clearly impractical to have a connection to each one. Instead, the desired bit has to be addressed before it can be read or written. The size of the chip package restricts the number of pins available, so that large memories use the same address pins more than once. The bits are arranged internally as rows and columns, and the row address and the column address are specified sequentially on the same pins.

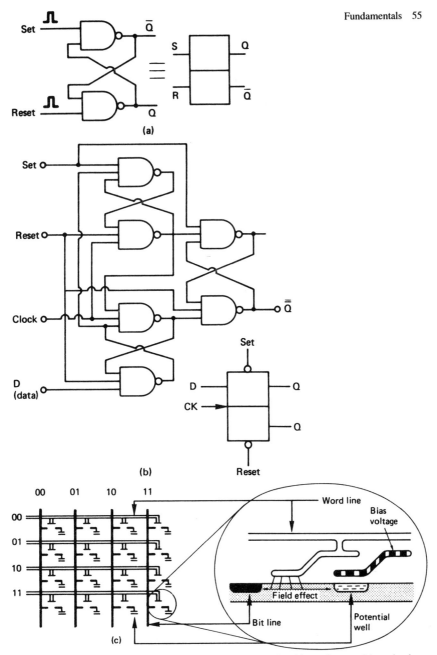

Figure 2.38 Digital semiconductor memory types. In (a), one data bit can be stored in a simple set-reset latch, which has little application because the D-type latch in (b) can store that state of the single data input when the clock occurs. These devices can be implemented with bipolar transistors of FETs, and are called static memories because they can store indefinitely. They consume a lot of power.

In (c), a bit is stored as the charge in a potential well in the substrate of a chip. It is accessed by connecting the bit line with the field effect from the word line. The single well where the two lines cross can then be written or read. These devices are called dynamic RAMs because the charge decays, and they must be read and rewritten (refreshed) periodically.

56 Fundamentals

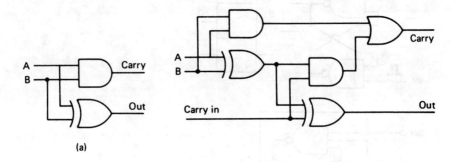

Data A	Bits B	Carry in	Out	Carry out
0	0	0	0	0
0	0	1	1	0
0	1	0	1	0
0	1	1	0	1
1	0	0	1	0
1	0	1	0	1
1	1	0	0	1
1	1	1	1	1

(b)

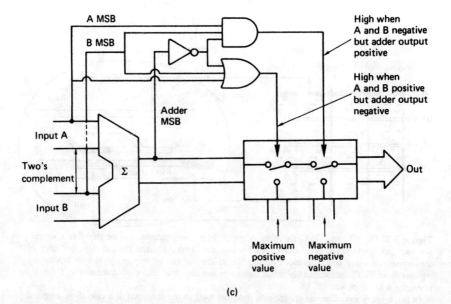

Figure 2.39 (a) Half adder; (b) full-adder circuit and truth table; (c) comparison of sign bits prevents wraparound on adder overflow by substituting clipping level.

2.21 Binary adding

The circuitry necessary for adding pure binary or two's complement numbers is shown in Figure 2.39. Addition in binary requires two bits to be taken at a time from the same position in each word, starting at the least significant bit. Should both be ones, the output is zero, and there is a *carry-out* generated. Such a circuit is called a half adder, shown in Figure 2.39(a), and is suitable for the least significant bit of the calculation. All higher stages will require a circuit which can accept a carry input as well as two data inputs. This is known as a full adder (Figure 2.39(b)). Multibit full adders are available in chip form, and have carry-in and carry-out terminals to allow them to be cascaded to operate on long wordlengths. Such a device is also convenient for inverting a two's complement number, in conjunction with a set of inverters. The adder chip has one set of inputs grounded, and the carry-in permanently held true, such that it adds 1 to the one's complement number from the inverter.

When mixing by adding sample values, care has to be taken to ensure that if the sum of the two sample values exceeds the number range the result will be clipping rather than wraparound. In two's complement, the action necessary depends on the polarities of the two signals. Clearly if one positive and one negative number are added, the result cannot exceed the number range. If two positive numbers are added, the symptom of positive overflow is that the most significant bit sets, causing an erroneous negative result, whereas a negative overflow results in the most significant bit clearing. The overflow control circuit will be designed to detect these two conditions, and override the adder output. If the MSB of both inputs is zero, the numbers are both positive; thus if the sum has the MSB set, the output is replaced with the maximum positive code (0111...). If the MSB of both inputs is set, the numbers are both negative, and if the sum has no MSB set, the output is replaced with the maximum negative code (1000...). These conditions can also be connected to warning indicators. Figure 2.39(c) shows this system in hardware. The resultant clipping on overload is sudden, and sometimes a PROM is included which translates values around and beyond maximum to soft-clipped values below or equal to maximum.

A storage element can be combined with an adder to obtain a number of useful functional blocks which will crop up frequently in digital equipment. Figure 2.40(a) shows that a latch is connected in a feedback loop around an adder. The latch contents are added to the input each time it is clocked. The configuration is known as an accumulator in computation because it adds up or accumulates values fed into it. In filtering, it is known as a discrete time integrator. If the input is held at some constant value, the output increases by that amount on each clock. The output is thus a sampled ramp.

Figure 2.40(b) shows that the addition of an inverter allows the difference between successive inputs to be obtained. This is digital differentiation. The output is proportional to the slope of the input.

2.22 Gain control by multiplication

Gain control is used extensively in compression systems. Digital filtering and transform calculations rely heavily on it, as do the requantizing processes which perform the actual compression. Gain is controlled in the digital domain by

58 Fundamentals

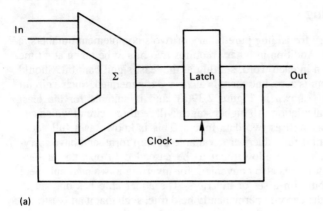

(a)

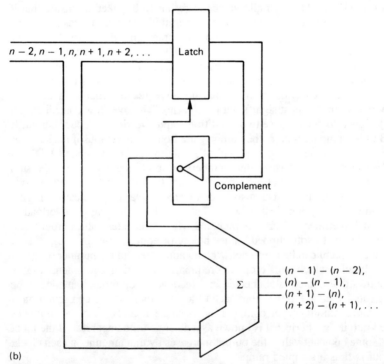

(b)

Figure 2.40 Two configurations which are common in processing. In (a) the feedback around the adder adds the previous sum to each input to perform accummulation or digital integration. In (b) an inverter allows the difference between successive inputs to be computed. This is differentiation.

multiplying each sample value by a coefficient. If that coefficient is less than 1 attenuation will result; if it is greater than 1, amplification can be obtained.

Multiplication in binary circuits is difficult. It can be performed by repeated adding, but this is too slow to be of any use. In fast multiplication, one of the inputs will be simultaneously multiplied by one, two, four, etc. by hard-wired bit

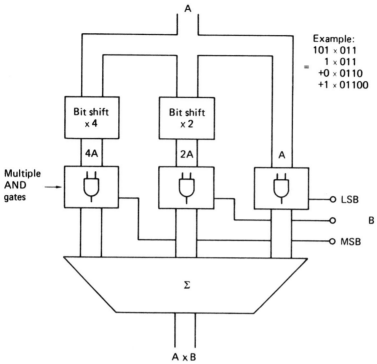

Figure 2.41 Structure of fast multiplier: the input A is multiplied by 1, 2, 4, 8, etc., by bit shifting. The digits of the B input then determine which multiples of A should be added together by enabling AND gates between the shifters and the adder. For long wordlengths, the number of gates required becomes enormous, and the device is best implemented in a chip.

shifting. Figure 2.41 shows that the other input bits will determine which of these powers will be added to produce the final sum, and which will be neglected. If multiplying by five, the process is the same as multiplying by four, multiplying by one, and adding the two products. This is achieved by adding the input to itself shifted two places. As the wordlength of such a device increases, the complexity increases exponentially, so this is a natural application for an integrated circuit.

References

1. WATKINSON, J.R., *The Art of Digital Audio*. Oxford: Focal Press (1994)
2. BETTS, J.A., *Signal Processing Modulation and Noise*, Chapter 6. Sevenoaks: Hodder and Stoughton (1970)
3. MEYER, J., Time correction of anti-aliasing filters used in digital audio systems. *J. Audio Eng. Soc.*, **32**, 132–137 (1984)
4. BLESSER, B., Advanced A/D conversion and filtering: data conversion. In B. A. Blesser, B. Locanthi and T. G. Stockham Jr (eds), *Digital Audio*, pp. 37–53. New York: Audio Engineering Society (1983)
5. LAGADEC, R., WEISS, D. and GREUTMANN, R., High-quality analog filters for digital audio. Presented at the 67th Audio Engineering Society Convention (New York, 1980), preprint 1707(B-4)
6. ISHIDA, Y. *et al.*, A PCM digital audio processor for home use VTRs. Presented at the 64th AES Convention (New York, 1979), preprint 1528

7. RUMSEY, F.J. and WATKINSON, J.R., *The Digital Interface Handbook*. Oxford: Focal Press (1995)
8. AES, AES recommended practice for professional digital audio applications employing pulse code modulation: preferred sampling frequencies. AES5-1984 (ANSI S4.28-1984), *J. Audio Eng. Soc.*, **32**, 781-785 (1984)
9. LIPSHITZ, S.P., WANNAMAKER, R.A. and VANDERKOOY, J., Quantization and dither: a theoretical survey. *J. Audio Eng. Soc.*, **40**, 355-375 (1992)
10. ROBERTS, L.G., Picture coding using pseudo-random noise. *IRE Trans. Inform. Theory*, **IT-8**, 145-154 (1962)
11. WATKINSON, J.R., *The Art of Digital Video*. Oxford: Focal Press (1994)

Chapter 3
Processing for compression

Virtually all compression systems rely on some combination of the basic processes outlined in this chapter. In order to understand compression a good grasp of filtering and transforms is essential along with motion estimation for video applications. These processes only express the information in the best way for the actual compression stage, which almost exclusively begins by using requantizing to shorten the wordlength. In this chapter the principles of filters and transforms will be explored, along with motion estimation and requantizing. These principles will be useful background for the next two chapters.

3.1 Filters

Filtering is inseparable from digital video and audio compression. There are many parallels between analog, digital and optical filters, which this section treats as a common subject. The main difference between analog and digital filters is that in the digital domain very complex architectures can be constructed at low cost in LSI and that arithmetic calculations are not subject to component tolerance or drift. Because of the sampled nature of the signal, whatever the response at low frequencies may be, all digital channels act as low-pass filters cutting off at the Nyquist limit, or half the sampling frequency.

Filtering may modify the frequency response of a system, and/or the phase response. Every combination of frequency and phase response determines the impulse response in the time domain. Figure 3.1 shows that impulse response testing tells a great deal about a filter. In a perfect filter, all frequencies should experience the same time delay. If some groups of frequencies experience a different delay from others, there is a group-delay error. As an impulse contains an infinite spectrum, a filter suffering from group-delay error will separate the different frequencies of an impulse along the time axis. A pure delay will cause a phase shift proportional to frequency, and a filter with this characteristic is said to be phase-linear. The impulse response of a phase-linear filter is symmetrical. If a filter suffers from group-delay error it cannot be phase-linear.

Filters can be described in two main classes, as shown in Figure 3.2, according to the nature of the impulse response. Finite-impulse response (FIR) filters are always stable and, as their name suggests, respond to an impulse once, as they have only a forward path. In the temporal domain, the time for which the filter responds to an input is finite, fixed and readily established. The same is therefore true about the distance over which an FIR filter responds in the spatial domain.

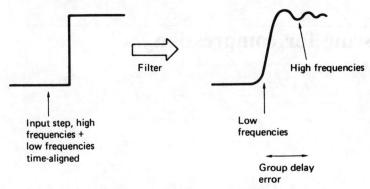

Figure 3.1 Group delay time-displaces signals as a function of frequency.

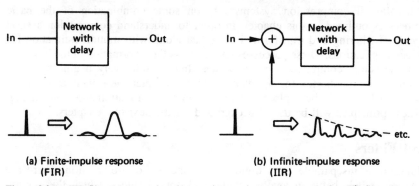

Figure 3.2 An FIR filter (a) responds only to an input, whereas the output of an IIR filter (b) continues indefinitely rather like a decaying echo.

FIR filters can be made perfectly phase linear if required. Most filters used for compression purposes fall into this category.

Infinite-impulse response (IIR) filters respond to an impulse indefinitely and are not necessarily stable, as they have a return path from the output to the input. For this reason they are also called recursive filters. As the impulse response is not symmetrical, IIR filters are not phase linear. Compression systems are not intended to change the phase characteristics of the signals they handle, and so linear phase filtering is essential and only FIR filters will be considered in detail here.

An FIR filter works by graphically constructing the impulse response for every input sample. It is first necessary to establish the correct impulse response. Figure 3.3(a) shows an example of a low-pass filter which cuts off at $\frac{1}{4}$ of the sampling rate. The impulse response of a perfect low-pass filter is a $\sin x/x$ curve, where the time between the two central zero crossings is the reciprocal of the cut-off frequency. According to the mathematics, the waveform has always existed, and carries on for ever. The peak value of the output coincides with the input impulse. This means that the filter is not causal, because the output has changed before the input is known. Thus in all practical applications it is necessary to truncate the extreme ends of the impulse response, which causes an aperture effect, and to introduce a time delay in the filter equal to half the duration of the truncated

Processing for compression 63

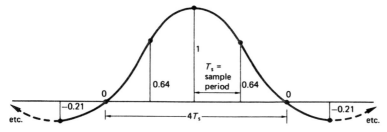

Figure 3.3(a) The impulse response of an LPF is a sin x/x curve which stretches from $-\infty$ to $+\infty$ in time. The ends of the response must be neglected, and a delay introduced to make the filter causal.

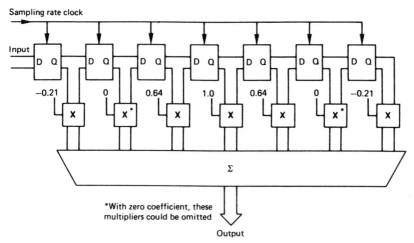

Figure 3.3(b) The structure of an FIR LPF, Input samples shift across the register and at each point are multiplied by different coefficients.

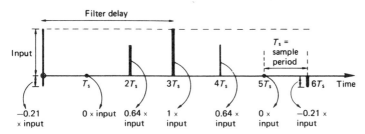

Figure 3.3(c) When a single unit sample shifts across the circuit of Figure 3.3(b), the impulse response is created at the output as the impulse is multiplied by each coefficient in turn.

impulse in order to make the filter causal. As an input impulse is shifted through the series of registers in Figure 3.3(b), the impulse response is created, because at each point it is multiplied by a coefficient as in Figure 3.3(c). These coefficients are simply the result of sampling and quantizing the desired impulse response. Clearly the sampling rate used to sample the impulse must be the same as the sampling rate for which the filter is being designed. In practice the coefficients are calculated, rather than attempting to sample an actual impulse response. The coefficient wordlength will be a compromise between cost and performance. An exception to this is in some video filters where the wordlength is short enough to allow the multipliers to be implemented as look-up tables.

Because the input sample shifts across the system registers to create the shape of the impulse response, the configuration is also known as a transversal filter. In operation with real sample streams, there will be several consecutive sample values in the filter registers at any time in order to convolve the input with the impulse response.

Simply truncating the impulse response causes an abrupt transition from input samples which matter and those which do not. Truncating the filter superimposes a rectangular shape on the time-domain impulse response. In the frequency domain the rectangular shape transforms to a $\sin x/x$ characteristic which is superimposed on the desired frequency response as a ripple. One consequence of this is known as Gibbs' phenomenon: a tendency for the response to peak just before the cut-off frequency.[1,2] As a result, the length of the impulse which must be considered will depend not only on the frequency response, but also on the amount of ripple which can be tolerated. If the relevant period of the impulse is measured in sample periods, the result will be the number of points or multiplications needed in the filter. Figure 3.4 compares the performance of filters with different numbers of points. Video filters may use as few as eight points whereas a high-quality digital audio FIR filter may need as many as 96 points.

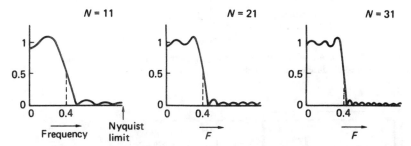

Figure 3.4 The truncation of the impulse in an FIR filter caused by the use of a finite number of points (N) results in ripple in the response. Shown here are three different numbers of points for the same impulse response. The filter is an LPF which rolls off at 0.4 of the fundamental interval. (Courtesy *Philips Technical Review*)

Rather than simply truncate the impulse response in time, it is better to make a smooth transition from samples which do not count to those that do. This can be done by multiplying the coefficients in the filter by a window function which peaks in the centre of the impulse. Figure 3.5 shows some different window functions and their responses. The rectangular window is the case of truncation, and the response is shown at I. A linear reduction in weight from the centre of the window to the edges characterizes the Bartlett window II, which trades ripple for an increase in transition-region width. At III is shown the Hanning window, which is essentially a raised cosine shape. Not shown is the similar Hamming

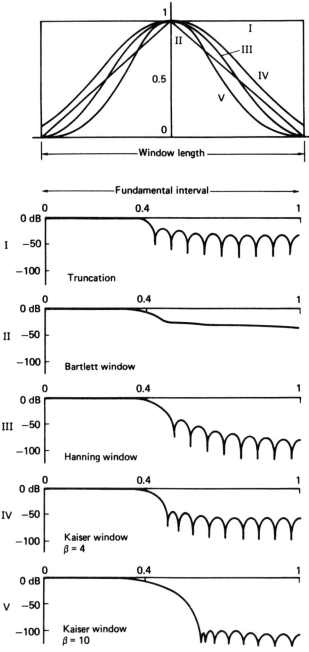

Figure 3.5 The effect of window functions. At top, various window functions are shown in continuous form. Once the number of samples in the window is established, the continuous functions shown here are sampled at the appropriate spacing to obtain window coefficients. These are multiplied by the truncated impulse response coefficients to obtain the actual coefficients used by the filter. The amplitude responses I–V correspond to the window functions illustrated. (Responses courtesy *Philips Technical Review*)

window, which offers a slightly different trade-off between ripple and the width of the main lobe. The Blackman window introduces an extra cosine term into the Hamming window at half the period of the main cosine period, reducing Gibbs' phenomenon and ripple level, but increasing the width of the transition region. The Kaiser window is a family of windows based on the Bessel function, allowing various trade-offs between ripple ratio and main lobe width. Two of these are shown at IV and V.

Filter coefficients can be optimized by computer simulation. One of the best-known techniques used is the Remez exchange algorithm, which converges on the optimum coefficients after a number of iterations.

In the example of Figure 3.6, the low-pass filter of Figure 3.3 is shown with a Bartlett window. Acceptable ripple determines the number of significant sample periods embraced by the impulse. This determines in turn both the number of points in the filter and the filter delay. As the impulse is symmetrical, the delay will be half the impulse period. The impulse response is a $\sin x/x$ function, and this has been calculated in the figure. The $\sin x/x$ response is next multiplied by the window function to give the windowed impulse response.

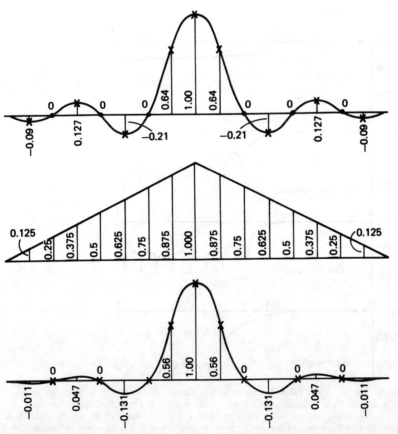

Figure 3.6 A truncated sin x/x impulse (top) is multiplied by a Bartlett window function (centre) to produce the actual coefficients used (bottom).

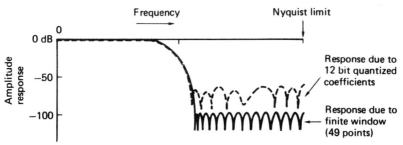

Figure 3.7 Frequency response of a 49 point transversal filter with infinite precision (solid line) shows ripple due to finite window size. Quantizing coefficients to 12 bits reduces attenuation in the stopband. (Responses courtesy *Philips Technical Review*)

If the coefficients are not quantized finely enough, it will be as if they had been calculated inaccurately, and the performance of the filter will be less than expected. Figure 3.7 shows an example of quantizing coefficients. Conversely, raising the wordlength of the coefficients increases cost.

The FIR structure is inherently phase linear because it is easy to make the impulse response absolutely symmetrical. Because of this inherent phase-linearity, an FIR filter can be designed for a specific impulse response, and the frequency response will follow.

The frequency response of the filter can be changed at will by changing the coefficients. A programmable filter only requires a series of PROMs to supply the coefficients; the address supplied to the PROMs will select the response. The frequency response of a digital filter will also change if the clock rate is changed, so it is often less ambiguous to specify a frequency of interest in a digital filter in terms of a fraction of the fundamental interval rather than in absolute terms. The configuration shown in Figure 3.3 serves to illustrate the principle. The units used on the diagrams are sample periods and the response is proportional to these periods or spacings, and so it is not necessary to use actual figures.

Where the impulse response is symmetrical, it is often possible to reduce the number of multiplications, because the same product can be used twice, at equal distances before and after the centre of the window. This is known as folding the filter. A folded filter is shown in Figure 3.8.

3.2 The quadrature mirror filter

Audio compression often uses a process known as bandsplitting which splits up the audio spectrum into a series of frequency ranges. Bandsplitting is complex and requires a lot of computation. One bandsplitting method which is useful is quadrature mirror filtering.[3] The QMF is a kind of twin FIR filter which converts a PCM sample stream into two sample streams of half the input sampling rate, so that the output data rate equals the input data rate. The frequencies in the lower half of the audio spectrum are carried in one sample stream, and the frequencies in the upper half of the spectrum are carried in the other. Whilst the lower-frequency output is a PCM band-limited representation of the input waveform, the upper frequency output isn't. A moment's thought will reveal that it could not be because the sampling rate is not high enough. In fact the upper half of the input spectrum has been heterodyned down to the same frequency band as the

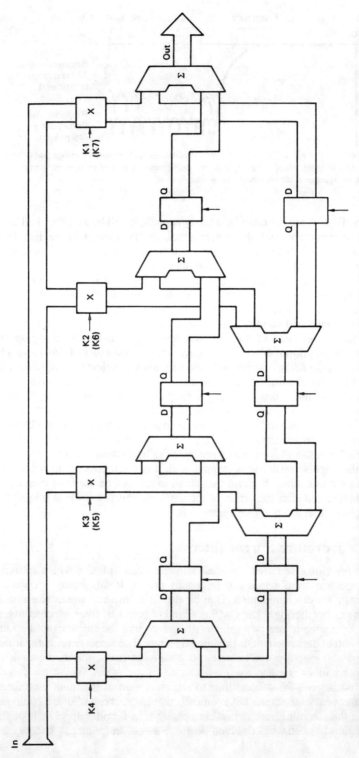

Figure 3.8 A seven-point folded filter for a symmetrical impulse reponse. In this case K1 and K7 will be identical, and so the input sample can be multipled once, and the product fed into the output shift system in two different places. The centre coefficient K4 appears once. In an even-numbered filter the centre coefficient would also be used twice.

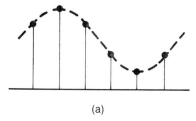 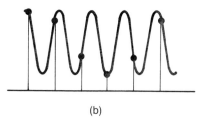

(a) (b)

Figure 3.9 The sample stream shown would ordinarily represent the waveform shown in (a), but if it is known that the original signal could exist only between two frequencies then the waveform in (b) must be the correct one. A suitable bandpass reconstruction filter, or synthesis filter, will produce the waveform in (b).

lower half by the clever use of aliasing. The waveform is unrecognizable, but when heterodyned back to its correct place in the spectrum in an inverse step, the correct waveform will result once more. Figure 3.9 shows how the idea works. Sampling theory states that the sampling rate needed must be at least twice the highest frequency in the signal to be sampled, but this is only true for a broadband signal. If the signal is band-limited, the sampling rate need only be more than twice the signal bandwidth. As the QMF bisects the frequency range of its input, the upper range can be expressed by half the previous sampling rate.

Downsampled signals of this kind can be reconstructed by an interpolator having a bandpass response, also known as a synthesis filter. As only signals within the passband can be output, it is clear from Figure 3.9 that the waveform which will result is the original as the intermediate aliased waveform lies outside the passband.

Figure 3.10 shows the operation of a simple QMF. In (a) the input spectrum of the PCM audio is shown, having an audio baseband extending up to half the sampling rate and the usual lower sideband extending down from there up to the sampling frequency. The input is passed through an FIR low-pass filter which cuts off at one-quarter of the sampling rate to give the spectrum shown in (b). The input also passes in parallel through a second FIR filter which is physically identical, but the coefficients are different. The impulse response of the FIR LPF is multiplied by a cosinusoidal waveform which amplitude-modulates it. The resultant impulse gives the filter a frequency response shown in (c). This is a mirror image of the LPF response. If certain criteria are met, the overall frequency response of the two filters is flat. The spectra of both (b) and (c) show that both are oversampled by a factor of 2 because they are half empty. As a result both can be decimated by a factor of two, which is the equivalent of dropping every other sample. In the case of the lower half of the spectrum, nothing remarkable happens. In the case of the upper half of the spectrum, it has been resampled at half the original frequency as shown in (d). The result is that the upper half of the audio spectrum aliases or heterodynes to the lower half.

An inverse QMF will recombine the bands into the original broadband signal. It is a feature of a QMF/inverse QMF pair that any energy near the band edge which appears in both bands due to inadequate selectivity in the filtering reappears at the correct frequency in the inverse filtering process provided that there is uniform quantizing in all of the sub-bands. In practical coders this

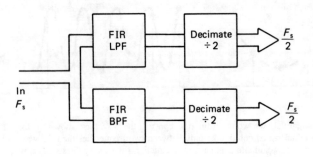

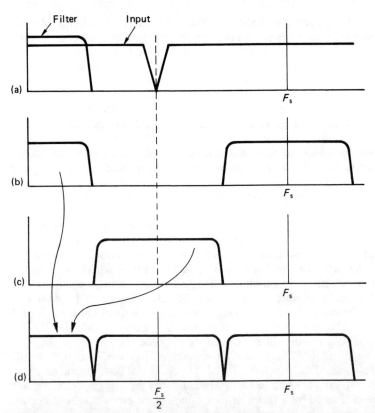

Figure 3.10 The quadrature mirror filter. In (a) the input spectrum has an audio baseband extending up to half the sampling rate. The input is passed through an FIR low-pass filter which cuts off at one-quarter of the sampling rate to give the spectrum shown in (b). The input also passes in parallel through a second FIR filter whose impulse response has been multiplied by a cosinusoidal waveform in order to amplitude-modulate it. The resultant impulse gives the filter a mirror image frequency response shown in (c). The spectra of both (b) and (c) show that both are oversampled by a factor of two because they are half empty. As a result both can be decimated by a factor of two, resulting in (d) in two identical Nyquist-sampled frequency bands of half the original width.

criterion is not met, but any residual artifacts are sufficiently small to be masked.

The audio band can be split into as many bands as required by cascading QMFs in a tree. However, each stage can only divide the input spectrum in half. In some coders certain sub-bands will have passed through one splitting stage more than others and will have half their bandwidth.[4] A delay is required in the wider sub-band data for time alignment.

A simple quadrature mirror is computationally intensive because sample values are calculated which are later decimated or discarded, and an alternative is to use polyphase pseudo-QMF filters[5] or wave filters[6] in which the filtering and decimation process is combined. Only wanted sample values are computed. In a polyphase filter a set of samples is shifted into position in the transversal register and then these are multiplied by different sets of coefficients and accumulated in each of several phases to give the value of a number of different samples between input samples. In a polyphase QMF, the same approach is used. Figure 3.11 shows an example of a 32 band polyphase QMF having a 512 sample window. With 32 sub-bands, each band will be decimated to $\frac{1}{32}$ of the input sampling rate. Thus only one sample in 32 will be retained after the combined filter/decimate operation. The polyphase QMF only computes the value of the sample which is to be retained in each sub-band. The filter works in 32 different phases with the same samples in the transversal register. In the first phase, the coefficients will describe the impulse response of a low-pass filter, the so-called prototype filter, and the result of 512 multiplications will be accumulated to give a single sample in the first band. In the second phase the coefficients will be obtained by multiplying the impulse response of the prototype filter by a cosinusoid at the centre frequency of the second band. Once more 512 multiply accumulates will be required to obtain a single sample in the second band. This is repeated for each of the 32 bands, and in each case a different centre frequency

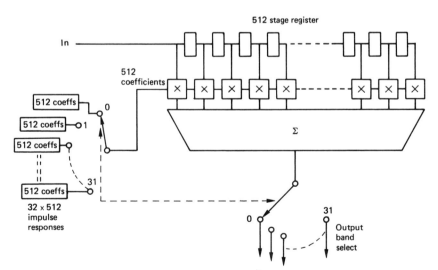

Figure 3.11 In polyphase QMF the same input samples are subject to computation using coefficient sets in many different time-multiplexed phases. The decimation is combined with the filtering so only wanted values are computed.

is obtained by multiplying the prototype impulse by a different modulating frequency. Following 32 such computations, 32 output samples, one in each band, will have been computed. The transversal register then shifts 32 samples and the process repeats.

The principle of the polyphase QMF is not so different from the techniques used to compute a frequency transform and effectively blurs the distinction between sub-band coding and transform coding.

3.3 Filtering for video noise reduction

The basic principle of all video noise reducers is that there is a certain amount of correlation between the video content of successive frames, whereas there is no correlation between the noise content.

A basic recursive device is shown in Figure 3.12. There is a frame store which acts as a delay, and the output of the delay can be fed back to the input through an attenuator, which in the digital domain will be a multiplier. In the case of a still picture, successive frames will be identical, and the recursion will be large. This means that the output video will actually be the average of many frames. If there is movement of the image, it will be necessary to reduce the amount of recursion to prevent the generation of trails or smears. Probably the most famous examples of recursion smear are the television pictures sent back of astronauts walking on the moon. The received pictures were very noisy and needed a lot of averaging to make them viewable. This was fine until the astronaut moved. The technology of the day did not permit motion sensing.

The noise reduction increases with the number of frames over which the noise is integrated, but image motion prevents simple combining of frames. If motion estimation is available, the image of a moving object in a particular frame can be integrated from the images in several frames which have been superimposed on

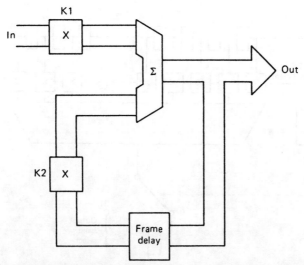

Figure 3.12 A basic recursive device feeds back the output to the input via a frame store which acts as a delay. The characteristics of the device are controlled totally by the values of the two coefficients K1 and K2 which control the multipliers.

the same part of the screen by displacements derived from the motion measurement. The result is that greater reduction of noise becomes possible.[7]

In a median filter, sample values adjacent to the one under examination are considered. These may be in the same place in previous or subsequent images, or nearby in the same image. A median filter computes the distribution of values on all of its input points. If the value of the centre point lies centrally within the distribution then it is considered to be valid and is passed to the output without change. In this case the median filter has no effect whatsoever. However, if the value of the centre point is at the edge of the distribution it is considered to be in error due to impulsive noise and a different input point, nearest the mean, is selected as the output pixel. Effectively a pixel value from nearby is used to conceal the error. The median filter is very effective against dropouts, film dirt and bit errors.

3.4 Transforms

At its simplest a transform is a process which takes information in one domain and expresses it in another. Audio and video signals are in the time domain: their voltages (or sample values) change as a function of time. Such signals are often transformed to the frequency domain for the purposes of compression. Whilst frequency in audio has traditionally meant temporal frequency measured in Hertz, frequency in optics can also be spatial and measured in lines per millimetre (mm^{-1}) or in television in cycles per picture height (cph) or width (cpw). In the frequency domain the signal has been converted to a spectrum; a table of the energy at different temporal or spatial frequencies. If the input is repeating, the spectrum will also be sampled, i.e. it will exist at discrete frequencies. In real programme material, all frequencies are seldom present together and so those which are absent need not be transmitted and a coding gain is obtained. On reception an inverse transform or synthesis process converts back from the frequency domain to the time or space domains. To take a simple analogy, a frequency transform of piano music effectively works out what frequencies are present as a function of time; the transform works out which notes were played and so the information is not really any different from that contained in the original sheet music.

One frequently encountered way of entering the frequency domain from the time or spatial domains is the Fourier transform or its equivalent in sampled systems, the discrete Fourier transform (DFT). Fourier analysis holds that any periodic waveform can be reproduced by adding together an arbitrary number of harmonically related sinusoids of various amplitudes and phases. Figure 3.13 shows how a square wave can be built up of harmonics. The spectrum can be drawn by plotting the amplitude of the harmonics against frequency. It will be seen that this gives a spectrum which is a decaying wave. It passes through zero at all even multiples of the fundamental. The shape of the spectrum is a $\sin x/x$ curve. If a square wave has a $\sin x/x$ spectrum, it follows that a filter with a rectangular impulse response will have a $\sin x/x$ frequency response. It will be recalled that an ideal low-pass filter has a rectangular spectrum, and this has a $\sin x/x$ impulse response. These characteristics are known as a transform pair. In transform pairs, if one domain has one shape of the pair, the other domain will have the other shape. Thus a square wave has a $\sin x/x$ spectrum and a $\sin x/x$ impulse has a square spectrum. Figure 3.14 shows a number of transform pairs.

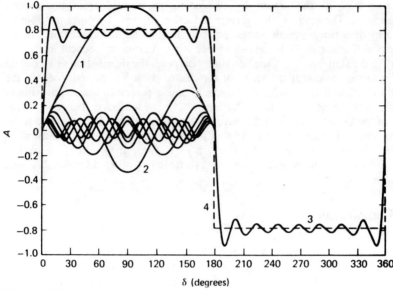

Figure 3.13 Fourier analysis of a square wave into fundamental and harmonics. A, amplitude; δ, phase of fundamental wave in degrees; 1, first harmonic (fundamental); 2, odd harmonics 3–15; 3, sum of harmonics 1–15; 4, ideal square wave.

Note the pulse pair. A time-domain pulse of infinitely short duration has a flat spectrum. Thus a flat waveform, i.e. a constant voltage, has only zero Hz in its spectrum. Interestingly, the transform of a Gaussian response is still Gaussian.

The Fourier transform specifies the amplitude and phase of the frequency components just once and such sine waves are endless. As a result the Fourier transform is only valid for periodic waveforms, i.e. those which repeat endlessly. Real programme material is not like that and so it is necessary to break up the continuous time domain using windows. Figure 3.15(a) shows how a block of time is cut from the continuous input. By wrapping it into a ring it can be made to appear like a continuous periodic waveform for which a single transform, known as the short-time Fourier transform (STFT), can be computed. Note that in the Fourier transform of a periodic waveform the frequency bands have constant width. The inverse transform produces just such an endless waveform, but a window is taken from it and used as part of the output waveform. Rectangular windows are used in video compression, but are not generally adequate for audio because the discontinuities at the boundaries are audible. This can be overcome by shaping and overlapping the windows so that a cross fade occurs at the boundaries between them as in Figure 3.15(b).

As has been mentioned, the theory of transforms assumes endless periodic waveforms. If an infinite length of waveform is available, spectral analysis can be performed to infinite resolution but as the size of the window reduces, so too does the resolution of the frequency analysis. Intuitively it is clear that discrimination between two adjacent frequencies is easier if more cycles of both are available. In sampled systems, reducing the window size reduces the number of samples and so must reduce the number of discrete frequencies in the

Processing for compression 75

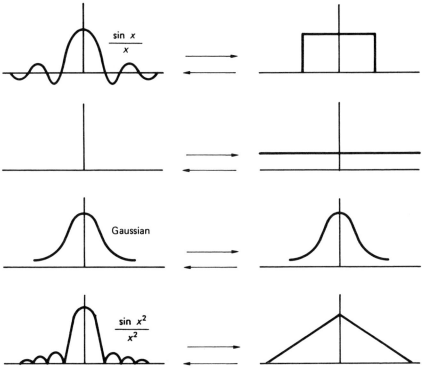

Figure 3.14 The concept of transform pairs illustrates the duality of the frequency (including spatial frequency) and time domains.

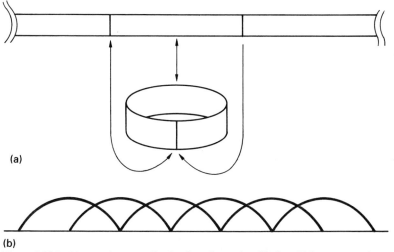

Figure 3.15 In (a) a continuous audio signal can be cut into blocks which are wrapped to make them appear periodic for the purposes of the Fourier transform. A better approach is to use overlapping windows to avoid discontinuities as in (b).

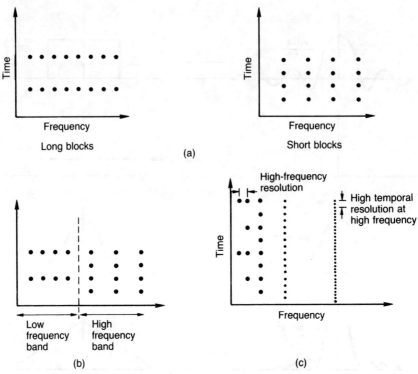

Figure 3.16 (a) In transforms greater certainty in the time domain leads to less certainty in the frequency domain and vice versa. Some transform coders split the spectrum as in (b) and use different window lengths in the two bands. In the recently developed wavelet transform the window length is inversely proportional to the frequency, giving the advantageous time/frequency characteristic shown in (c).

transform. Thus for good frequency resolution the window should be as large as possible. However, with large windows the time between updates of the spectrum is longer and so it is harder to locate events on the time axis. Figure 3.16(a) shows the effect of two window sizes in a conventional STFT and illustrates the principle of *uncertainty*, also known as the Heisenberg inequality.

According to the uncertainty theory one can trade off time resolution against frequency resolution. In most programme material, the time resolution required falls with frequency whereas the time (or spatial) resolution required rises with frequency. Fourier-based compression systems using transforms sometimes split the signal into a number of frequency bands in which different window sizes are available as in Figure 3.16(b). Some have variable-length windows which are selected according to the programme material. The Sony ATRAC system of the MiniDisc (see Chapter 4) uses these principles. Stationary material such as steady tones is transformed with long windows whereas transients are transformed with short windows.

The recently developed wavelet transform is one in which the window length is inversely proportional to the frequency. This automatically gives the advantageous time/frequency resolution characteristic shown in Figure 3.16(c). The wavelet transform is considered in more detail in Section 3.7.

Although compression uses transforms, the transform itself does not result in any data reduction, as there are usually as many coefficients as input samples. Paradoxically the transform increases the amount of data because the coefficient multiplications result in wordlength extension. Thus it is incorrect to refer to transform compression; instead the term transform-*based* compression should be used.

3.5 The Fourier transform

Figure 3.17 shows that if the amplitude and phase of each frequency component is known, linearly adding the resultant components in an inverse transform results in the original waveform. In digital systems the waveform is expressed as a number of discrete samples. As a result the Fourier transform analyses the

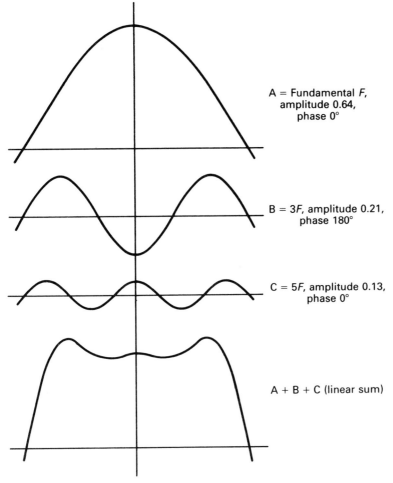

Figure 3.17 Fourier analysis allows the synthesis of any periodic waveform by the addition of discrete frequencies of appropriate amplitude and phase.

78 Processing for compression

signal into an equal number of discrete frequencies. This is known as a discrete Fourier transform or DFT, in which the number of frequency coefficients is equal to the number of input samples. The fast Fourier transform is no more than an efficient way of computing the DFT.[8] As was seen in the previous section, practical systems must use windowing to create short-term transforms.

It will be evident from Figure 3.17 that the knowledge of the phase of the frequency component is vital, as changing the phase of any component will seriously alter the reconstructed waveform. Thus the DFT must accurately analyse the phase of the signal components.

There are a number of ways of expressing phase. Figure 3.18 shows a point which is rotating about a fixed axis at constant speed. Looked at from the side, the point oscillates up and down at constant frequency. The waveform of that

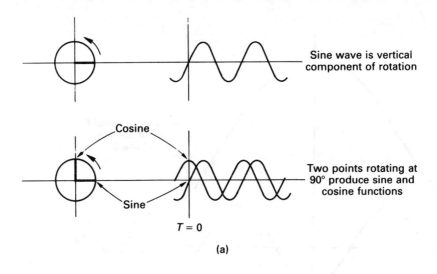

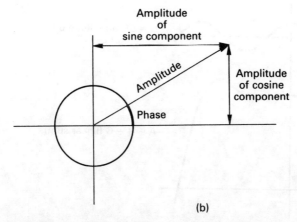

Figure 3.18 The origin of sine and cosine waves is to take a particular viewpoint of a rotation. Any phase can be synthesized by adding proportions of sine and cosine waves.

motion is a sine wave, and that is what we would see if the rotating point were to translate along its axis whilst we continued to look from the side.

One way of defining the phase of a waveform is to specify the angle through which the point has rotated at time zero ($T = 0$). If a second point is made to revolve at 90° to the first, it would produce a cosine wave when translated. It is possible to produce a waveform having arbitrary phase by adding together the sine and cosine wave in various proportions and polarities. For example, adding the sine and cosine waves in equal proportion results in a waveform lagging the sine wave by 45°.

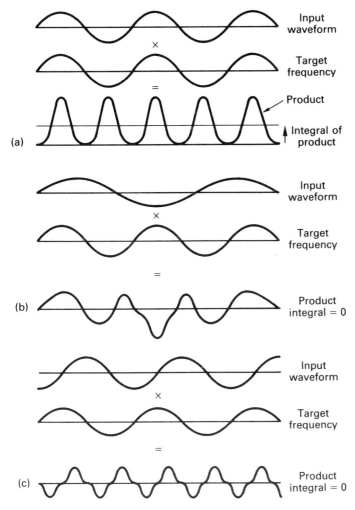

Figure 3.19 The input waveform is multiplied by the target frequency and the result is averaged or integrated. In (a) the target frequency is present and a large integral results. With another input frequency the integral is zero as in (b). The correct frequency will also result in a zero integral shown in (c) if it is at 90° to the phase of the search frequency. This is overcome by making two searches in quadrature.

80 Processing for compression

Figure 3.18 shows that the proportions necessary are respectively the sine and the cosine of the phase angle. Thus the two methods of describing phase can be readily interchanged.

The discrete Fourier transform spectrum analyses a string of samples by searching separately for each discrete target frequency. It does this by multiplying the input waveform by a sine wave, known as the basis function, having the target frequency and adding up or integrating the products. Figure 3.19(a) shows that multiplying by basis functions gives a non-zero integral when

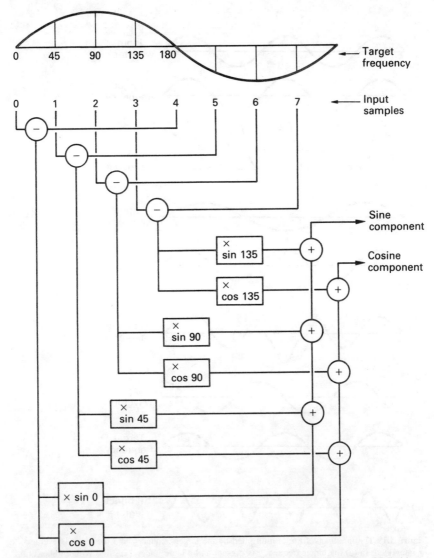

Figure 3.20 An example of a filtering search. Pairs of samples are subtracted and multiplied by sampled sine and cosine waves. The products are added to give the sine and cosine components of the search frequency.

Processing for compression 81

the input frequency is the same, whereas Figure 3.19(b) shows that with a different input frequency (in fact all other different frequencies) the integral is zero, showing that no component of the target frequency exists. Thus from a real waveform containing many frequencies all frequencies except the target frequency are excluded. The magnitude of the integral is proportional to the amplitude of the target component.

Figure 3.19(c) shows that the target frequency will not be detected if it is phase shifted 90° as the product of quadrature waveforms is always zero. Thus the discrete Fourier transform must make a further search for the target frequency using a cosine basis function. It follows from the arguments above that the relative proportions of the sine and cosine integrals reveal the phase of the input

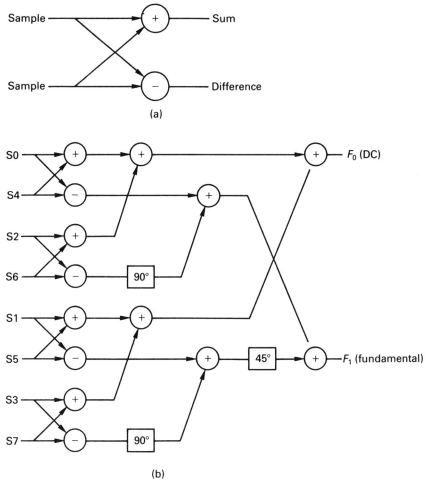

Figure 3.21 The basic element of an FFT is known as a butterfly as in (a) because of the shape of the signal paths in a sum and difference system. The use of butterflies to compute the first two coefficients is shown in (b). An actual example is given in (c) which should be compared with the result of (d) with a quadrature input. In (e) the butterflies for the first two coefficients form the basis of the computation of the third coefficient.

82 Processing for compression

component. Thus each discrete frequency in the spectrum must be the result of a pair of quadrature searches.

Searching for one frequency at a time as above will result in a DFT, but only after considerable computation. However, a lot of the calculations are repeated many times over in different searches. The fast Fourier transform gives the same result with less computation by logically gathering together all of the places where the same calculation is needed and making the calculation once.

The amount of computation can be reduced by performing the sine and cosine component searches together. Another saving is obtained by noting that every 180° the sine and cosine have the same magnitude but are simply inverted in sign. Instead of performing four multiplications on two samples 180° apart and adding the pairs of products, it is more economical to subtract the sample values and multiply twice, once by a sine value and once by a cosine value.

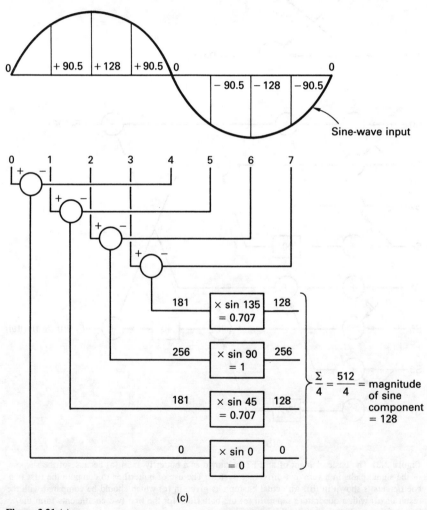

Figure 3.21 (c)

Processing for compression 83

The first coefficient is the arithmetic mean, which is the sum of all of the sample values in the block divided by the number of samples. Figure 3.20 shows how the search for the lowest frequency in a block is performed. Pairs of samples are subtracted as shown, and each difference is then multiplied by the sine and the cosine of the search frequency. The process shifts one sample period, and a new sample pair are subtracted and multiplied by new sine and cosine factors. This is repeated until all of the sample pairs have been multiplied. The sine and cosine products are then added to give the value of the sine and cosine coefficients respectively.

It is possible to combine the calculation of the DC component which requires the sum of samples and the calculation of the fundamental which requires sample differences by combining stages shown in Figure 3.21(a) which take a pair of samples and add and subtract them. Such a stage is called a butterfly because of

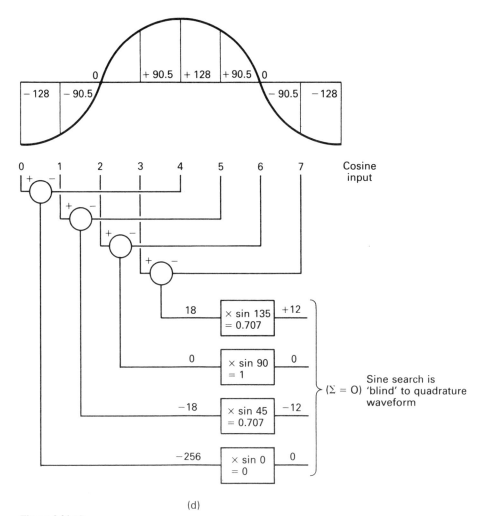

(d)

Figure 3.21 (d)

84 Processing for compression

the shape of the schematic. Figure 3.21(b) shows how the first two components are calculated. The phase rotation boxes attribute the input to the sine or cosine component outputs according to the phase angle. As shown, the box labelled 90° attributes nothing to the sine output, but unity gain to the cosine output. The 45° box attributes the input equally to both components.

Figure 3.21(c) shows a numerical example. If a sine wave input is considered where 0° coincides with the first sample, this will produce a zero sine coefficient and non-zero cosine coefficient. Figure 3.21(d) shows the same input waveform shifted by 90°. Note how the coefficients change over.

Figure 3.21(e) shows how the next frequency coefficient is computed. Note that exactly the same first-stage butterfly outputs are used, reducing the computation needed.

A similar process may be followed to obtain the sine and cosine coefficients of the remaining frequencies. The full FFT diagram for eight samples is shown in Figure 3.22(a). The spectrum this calculates is shown in Figure 3.22(b). Note that only half of the coefficients are useful in a real band-limited system because the remaining coefficients represent frequencies above one-half of the sampling rate.

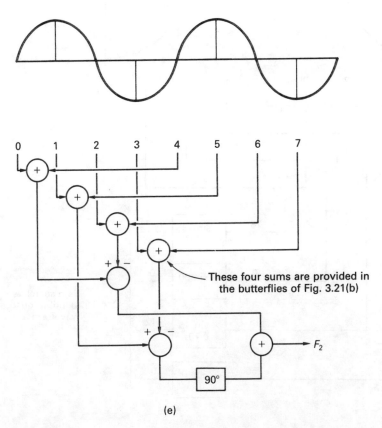

Figure 3.21 (e)

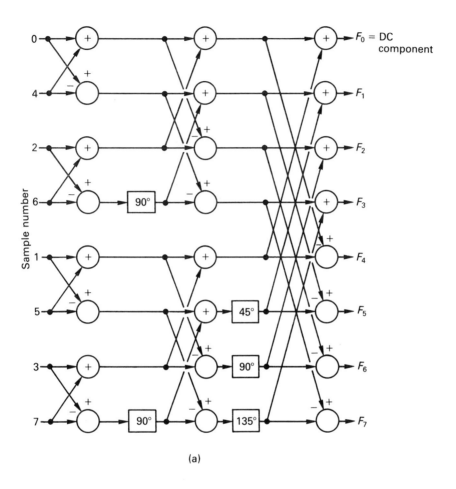

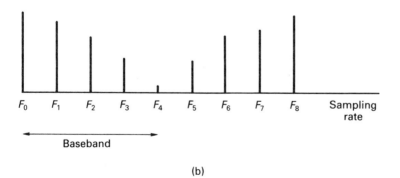

Figure 3.22 In (a) is the full butterfly diagram for an FFT. The spectrum this computes is shown in (b).

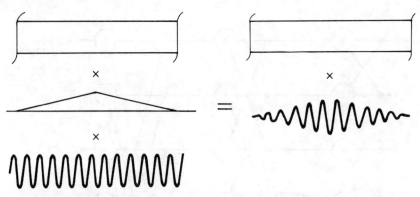

Figure 3.23 Multiplication of a windowed block by a sine wave basis function is the same as multiplying the raw data by a windowed basis function but requires less multiplication as the basis function is constant and can be pre-computed.

In STFTs the overlapping input sample blocks must be multiplied by window functions. The principle is the same as for the application in FIR filters shown in Section 3.1. Figure 3.23 shows that multiplying the search frequency by the window has exactly the same result except that this need be done only once and much computation is saved. Thus in the STFT the basis function is a windowed sine or cosine wave.

The FFT is used extensively in such applications as phase correlation, where the accuracy with which the phase of signal components can be analysed is essential. It also forms the foundation of the discrete cosine transform.

3.6 The discrete cosine transform (DCT)

The DCT is a special case of a discrete Fourier transform in which the sine components of the coefficients have been eliminated leaving a single number. This is actually quite easy. Figure 3.24(a) shows a block of input samples to a transform process. By repeating the samples in a time-reversed order and performing a discrete Fourier transform on the double-length sample set a DCT is obtained. The effect of mirroring the input waveform is to turn it into an even function whose sine coefficients are all zero. The result can be understood by considering the effect of individually transforming the input block and the reversed block. Figure 3.24(b) shows that the phase of all the components of one block are in the opposite sense to those in the other. This means that when the components are added to give the transform of the double length block all of the sine components cancel out, leaving only the cosine coefficients, hence the name of the transform.[9] In practice the sine component calculation is eliminated. Another advantage is that doubling the block length by mirroring doubles the frequency resolution, so that twice as many useful coefficients are produced. In fact a DCT produces as many useful coefficients as input samples. Clearly when the inverse transform is performed the reversed part of the waveform is discarded.

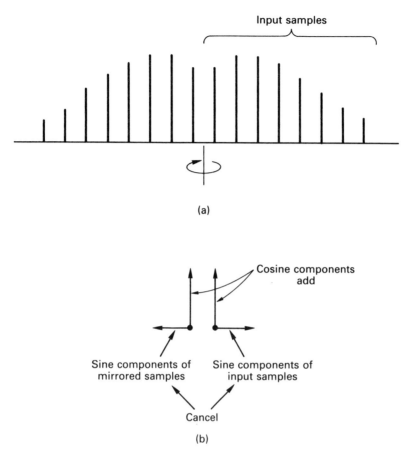

Figure 3.24 The DCT is obtained by mirroring the input block as shown in (a) prior to an FFT. The mirroring cancels out the sine components as in (b), leaving only cosine coefficients.

For image processing two-dimensional transforms are needed. In this case for every horizontal frequency, a search is made for all possible vertical frequencies. A two-dimensional DCT is shown in Figure 3.25. The DCT is separable in that the two-dimensional DCT can be obtained by computing in each dimension separately. Fast DCT algorithms are available.[10]

Figure 3.26 shows how a two-dimensional DCT is calculated by multiplying each pixel in the input block by terms which represent sampled cosine waves of various spatial frequencies. A given DCT coefficient is obtained when the result of multiplying every input pixel in the block is summed. Although most compression systems, including JPEG and MPEG, use square DCT blocks, this is not a necessity and rectangular DCT blocks are possible and are used in, for example, Digital Betacam.

The DCT is primarily used in data reduction processing because it converts the input waveform into a form where redundancy can be easily detected and removed.

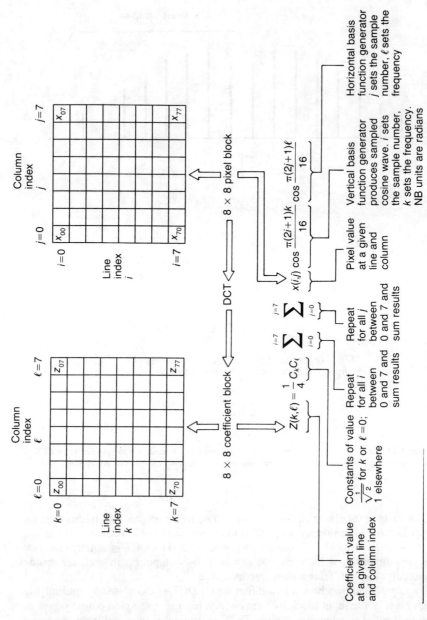

Figure 3.25 A two-dimensional DCT is calculated as shown here. Starting with an input pixel block one calculation is necessary to find a value for each coefficient. After 64 calculations using different basis functions the coefficient block is complete.

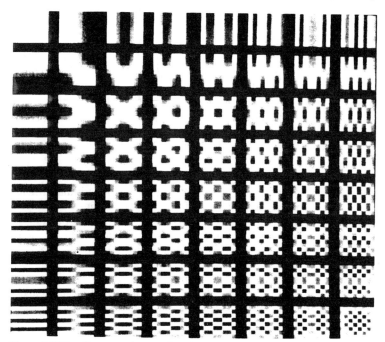

Figure 3.26 The discrete cosine transform breaks up an image area into discrete frequencies in two dimensions. The lowest frequency can be seen here at the top left corner. Horizontal frequency increases to the right and vertical frequency increases downwards.

3.7 The wavelet transform

The wavelet transform was not discovered by any one individual, but has evolved via a number of similar ideas and was only given a strong mathematical foundation relatively recently.[11,12] The wavelet transform is similar to the Fourier transform in that it has basis functions of various frequencies which are multiplied by the input waveform to identify the frequencies it contains. However, the Fourier transform is based on periodic signals and endless basis functions and requires windowing. The wavelet transform is fundamentally windowed, as the basis functions employed are not endless sine waves, but are finite on the time axis; hence the name. Wavelet transforms do not use a fixed window, but instead the window period is inversely proportional to the frequency being analysed. As a result a useful combination of time and frequency resolutions is obtained. High frequencies corresponding to transients in audio or edges in video are transformed with short basis functions and therefore are accurately located. Low frequencies are transformed with long basis functions which have good frequency resolution.

Figure 3.27 shows that that a set of wavelets or basis functions can be obtained simply by scaling (stretching or shrinking) a single wavelet on the time axis. Each wavelet contains the same number of cycles such that as the frequency reduces the wavelet gets longer. Thus the frequency discrimination of the wavelet transform is a constant fraction of the signal frequency. In a filter bank such a characteristic would be described as 'constant Q'. Figure 3.28 shows that the

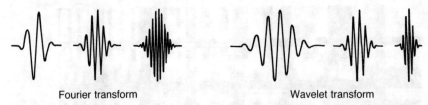

Fourier transform   Wavelet transform

Figure 3.27 Unlike discrete Fourier transforms, wavelet basis functions are scaled so that they contain the same number of cycles irrespective of frequency. As a result their frequency discrimination ability is a constant proportion of the centre frequency.

division of the frequency domain by a wavelet transform is logarithmic whereas in the Fourier transform the division is uniform. The logarithmic coverage is effectively dividing the frequency domain into octaves and as such parallels the frequency discrimination of human hearing. For a comprehensive introduction to wavelets having exhaustive references, the reader is referred to Rioul and Vetterli.[13]

As it is relatively recent, the wavelet transform has yet to be widely used although it shows great promise. It has been successfully used in audio and in commercially available non-linear video editors and in other fields such as radiology and geology.

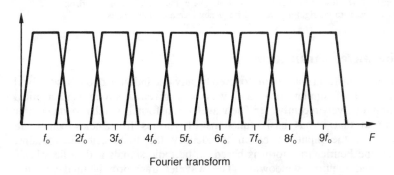

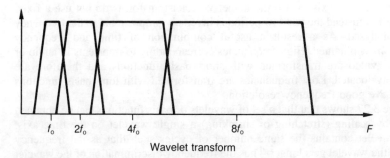

Figure 3.28 Wavelet transforms divide the frequency domain into octaves instead of the equal bands of the Fourier transform.

In video, wavelet compression does not display the 'blocking' of DCT-based coders at high compression factors; instead compression error is spread over the spectrum and appears as white noise.[14] It is naturally a multi-resolution transform allowing scalable decoding.

3.8 Motion compensation

Video data reduction relies on removing redundancy from the source signal. Whilst redundancy can be removed from individual images, it will be seen that higher compression factors are only achieved by eliminating data which are common in successive images, so that only data which are different need to be sent. Motion causes the image to move with respect to the sampling grid, causing all the sample values in a moving area to change and preventing effective reduction. With motion estimation the image movement can be cancelled because the comparison can be made along the motion axis rather than the time axis. Greater reduction factors are then possible because it is only necessary to send the motion parameters and a small number of genuine image differences.

Noise reduction in video signals works by combining together successive frames on the time axis such that the image content of the signal reinforces strongly whereas the random element in the signal due to noise does not. The noise reduction increases with the number of frames over which the noise is integrated, but image motion prevents simple combining of frames. If motion estimation is available, the image of a moving object in a particular frame can be integrated from the images in several frames which have been superimposed on the same part of the screen by displacements derived from the motion measurement. The result is that greater reduction of noise becomes possible.

3.9 Motion estimation techniques

There are three main methods of motion estimation which are to be found in various applications: block matching, gradient matching and phase correlation. Each have their own characteristics which are quite different.

3.9.1 Block matching

This is the simplest technique to follow. In a given picture, a block of pixels is selected and stored as a reference. If the selected block is part of a moving object, a similar block of pixels will exist in the next picture, but not in the same place. As Figure 3.29 shows, block matching simply moves the reference block around over the second picture looking for matching pixel values. When a match is found, the displacement needed to obtain it is used as a basis for a motion vector.

Whilst simple in concept, block matching requires an enormous amount of computation because every possible motion must be tested over the assumed range. Thus if the object is assumed to have moved over a 16 pixel range, then it will be necessary to test 16 different horizontal displacements in each of 16 vertical positions; in excess of 65 000 positions. At each position every pixel in the block must be compared with every pixel in the second picture. In typical video, displacements of twice the figure quoted here may be found, particularly in sporting events, and the computation then required becomes enormous.

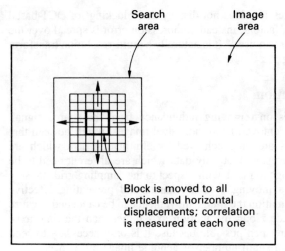

Figure 3.29 In block matching the search block has to be positioned at all possible relative motions within the search area and a correlation measured at each one.

One way of reducing the amount of computation is to perform the matching in stages where the first stage is inaccurate but covers a large motion range but the last stage is accurate but covers a small range.[15] The first matching stage is performed on a heavily filtered and subsampled picture, which contains far fewer pixels. When a match is found, the displacement is used as a basis for a second stage which is performed with a less heavily filtered picture. Eventually the last stage takes place to any desired accuracy. This hierarchical approach does reduce the computation required, but it suffers from the problem that the filtering of the first stage may make small objects disappear and they can never be found by subsequent stages if they are moving with respect to their background. Many televised sports events contain small, fast-moving objects. As the matching process depends upon finding similar luminance values, this can be confused by objects moving into shade or fades.

The simple block-matching systems described above can only measure motion to the nearest pixel. If more accuracy is required, interpolators will be needed to shift the image by subpixel distances before attempting a match. The complexity rises once more. In compression systems an accuracy of half a pixel is accurate enough for most purposes.

3.9.2 Gradient matching

At some point in a picture, the function of brightness with respect to distance across the screen will have a certain slope, known as the spatial luminance gradient. If the associated picture area is moving, the slope will traverse a fixed point on the screen and the result will be that the brightness now changes with respect to time. This is a temporal luminance gradient. Figure 3.30 shows the principle. For a given spatial gradient, the temporal gradient becomes steeper as the speed of movement increases. Thus motion speed can be estimated from the ratio of the spatial and temporal gradients.[16]

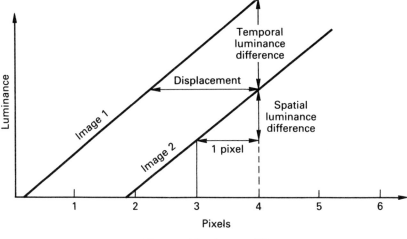

$$\text{Displacement (in pixels)} = \frac{\text{Temporal luminance difference}}{\text{Spatial luminance difference}}$$

Figure 3.30 The principle of gradient matching. The luminance gradient across the screen is compared with that through time.

The method only works well if the gradient remains essentially constant over the displacement distance; a characteristic which is not necessarily present in real video. In practice there are numerous processes which can change the luminance gradient. When an object moves so as to obscure or reveal the background, the spatial gradient will change from field to field even if the motion is constant. Variations in illumination, such as when an object moves into shade, also cause difficulty.

The process can be assisted by recursion, in which the motion is estimated over a larger number of fields, but this will result in problems at cuts.

3.9.3 Phase correlation

Phase correlation works by performing a discrete Fourier transform on two successive fields and then subtracting all of the phases of the spectral components. The phase differences are then subject to a reverse transform which directly reveals peaks whose positions correspond to motions between the fields.[15,17] The nature of the transform domain means that if the distance and direction of the motion is measured accurately, the area of the screen in which it took place is not. Thus in practical systems the phase correlation stage is followed by a matching stage not dissimilar to the block matching process. However, the matching process is steered by the motions from the phase correlation, and so there is no need to attempt to match at all possible motions. By attempting matching on measured motion the overall process is made much more efficient.

One way of considering phase correlation is that by using the Fourier transform to break the picture into its constituent spatial frequencies the hierarchical structure of block matching at various resolutions is in fact

94 Processing for compression

performed in parallel. In this way small objects are not missed because they will generate high-frequency components in the transform.

Although the matching process is simplified by adopting phase correlation, the Fourier transforms themselves require complex calculations. The high performance of phase correlation would remain academic if it were too complex to put into practice. However, if realistic values are used for the motion speeds which can be handled, the computation required by block matching actually exceeds that required for phase correlation.

The elimination of amplitude information from the phase correlation process ensures that motion estimation continues to work in the case of fades, objects moving into shade or flashguns firing.

The details of the Fourier transform have been described in Section 3.5. A one-dimensional example of phase correlation will be given here by way of introduction. A line of luminance, which in the digital domain consists of a series of samples, is a function of brightness with respect to distance across the screen. The Fourier transform converts this function into a spectrum of spatial frequencies (units of cycles per picture width) and phases.

All television signals must be handled in linear-phase systems. A linear-phase system is one in which the delay experienced is the same for all frequencies. If video signals pass through a device which does not exhibit linear phase, the various frequency components of edges become displaced across the screen. Figure 3.31 shows what phase linearity means. If the left-hand end of the frequency axis (zero Hz) is considered to be firmly anchored, but the right-hand end can be rotated to represent a change of position across the screen, it will be seen that as the axis twists evenly the result is phase shift proportional to frequency. A system having this characteristic is said to display linear phase.

In the spatial domain, a phase shift corresponds to a physical movement. Figure 3.32 shows that if between fields a waveform moves along the line, the lowest frequency in the Fourier transform will suffer a given phase shift, twice that frequency will suffer twice that phase shift and so on. Thus it is potentially possible to measure movement between two successive fields if the phase

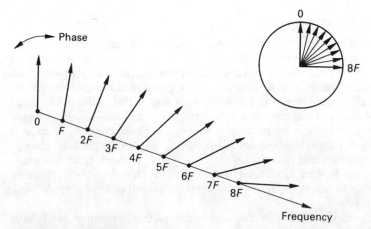

Figure 3.31 The definition of phase linearity is that phase shift is proportional to frequency. In phase-linear systems the waveform is preserved, and simply moves in time or space.

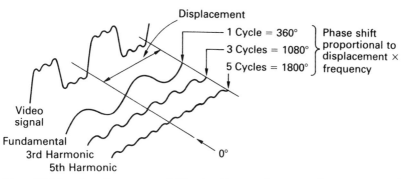

Figure 3.32 In a phase-linear system, shifting the video waveform across the screen causes phase shifts in each proportional to frequency.

differences between the Fourier spectra are analysed. This is the basis of phase correlation.

Figure 3.33 shows how a one-dimensional phase correlator works. The Fourier transforms of two lines from successive fields are computed and expressed in polar (amplitude and phase) notation (see Section 3.5). The phases of one transform are all subtracted from the phases of the same frequencies in the other transform. Any frequency component having significant amplitude is then normalized, or boosted to full amplitude.

The result is a set of frequency components which all have the same amplitude, but have phases corresponding to the difference between two fields. These coefficients form the input to an inverse transform. Figure 3.34(a) shows what happens. If the two fields are the same, there are no phase differences between the two, and so all of the frequency components are added with 0° phase to produce a single peak in the centre of the inverse transform. If, however, there was motion between the two fields, such as a pan, all of the components will have phase differences, and this results in a peak shown in Figure 3.34(b) which is displaced from the centre of the inverse transform by the distance moved. Phase correlation thus actually measures the movement between fields.

In the case where the line of video in question intersects objects moving at different speeds, Figure 3.34(c) shows that the inverse transform would contain one peak corresponding to the distance moved by each object.

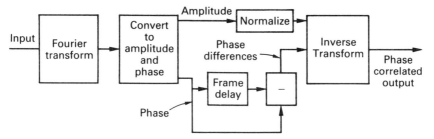

Figure 3.33 The basic components of a phase correlator.

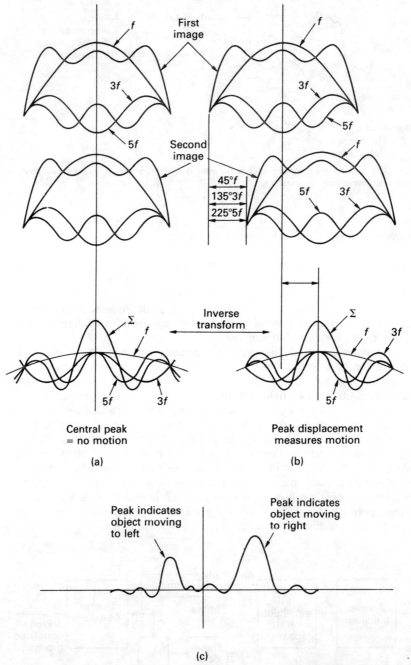

Figure 3.34 (a) The peak in the inverse transform is central for no motion. (b) In the case of motion, the peak shifts by the distance moved. (c) If there are several motions, each one results in a peak.

Processing for compression 97

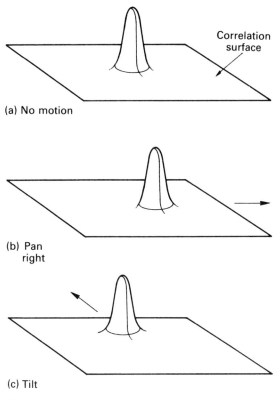

Figure 3.35 (a) A two-dimensional correlation surface has a central peak when there is no motion. (b) In the case of a pan, the peak moves laterally. (c) A camera tilt moves the peak at right angles to the pan.

Whilst this explanation has used one dimension for simplicity, in practice the entire process is two dimensional. A two-dimensional Fourier transform of each field is computed, the phases are subtracted, and an inverse two-dimensional transform is computed, the output of which is a flat plane out of which three-dimensional peaks rise. This is known as a correlation surface.

Figure 3.35(a) shows some examples of a correlation surface. In (a) there has been no motion between fields and so there is a single central peak. In (b) there has been a pan and the peak moves across the surface. In (c) the camera has been depressed and the peak moves upwards.

Where more complex motions are involved, perhaps with several objects moving in different directions and/or at different speeds, one peak will appear in the correlation surface for each object.

It is a fundamental strength of phase correlation that it actually measures the direction and speed of moving objects rather than estimating, extrapolating or searching for them. The motion can be measured to subpixel accuracy without excessive complexity.

However, the consequences of uncertainty are that accuracy in the transform domain is incompatible with accuracy in the spatial domain. Although phase correlation accurately measures motion speeds and directions, it cannot specify

3.10 Compression and requantizing

In compression systems the goal is to achieve coding gain by using fewer bits to represent the same information. The bandsplitting filters and transform techniques described earlier in this chapter *do not* achieve any coding gain. Their job is to express the information in a form in which redundancy can be identified. Paradoxically the output of a transform or a filter actually has a longer wordlength than the input because the integer input samples are multiplied by

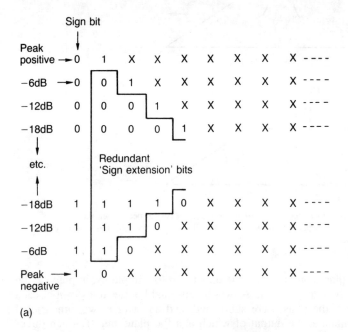

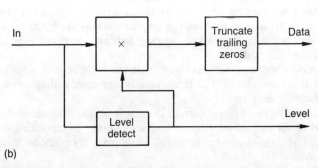

Figure 3.36 In two's complement coding, sign extension takes place as the level falls. These sign-extended bits are redundant and can be eliminated by multiplying by a level-dependent factor and neglecting the trailing zeros as shown in (b).

fractions. These processes have actually increased the redundancy in the signal. This section is concerned with the subsequent stage where the compression actually takes place.

Coding gain is obtained in one simple way: by shortening the wordlength of data words so that fewer bits are needed. These data words may be waveform samples in a sub-band-based system or coefficients in a transform-based system. In both cases the data will be expressed in two's complement form so that positive and negative values can be handled. Figure 3.36(a) shows various signal levels in two's complement coding. As the level falls a phenomenon called *sign extension* takes place where more and more bits at the most significant end of the word simply copy the sign bit (which is the MSB). Coding gain can be obtained by eliminating the redundant sign extension bits, as Figure 3.36(b) shows.

Taking bits out of the middle of a word is not straightforward and in practice the solution is to multiply by a level-dependent factor to eliminate the sign extension bits. If this is a power of 2 the useful bits will simply shift left up to the sign bit. The right-hand zeros are simply omitted from the transmission. On decoding, a compensating division must be performed. The multiplication factor must be transmitted along with the compressed data so this can be done. Clearly if only the sign extension bits are eliminated this process is lossless because exactly the same data values are available at the decoder.

The reason for sub-band filtering and transform coding now becomes clear because in real signals the levels in most sub-bands and the value of most coefficients are considerably less than the highest level.

In many cases the coding gain obtained in this way will not be enough and the wordlength has to be shortened even more. Following the multiplication, a larger number of bits are rounded off from the least significant end of the word. The result is that the same signal range is retained but it is expressed with less accuracy. It is as if the original analog signal had been converted using fewer quantizing steps, hence the term requantizing.

During the decoding process an *inverse quantizer* will be employed to convert the compressed value back to its original form. Figure 3.37 shows a requantizer and an inverse quantizer. The inverse quantizer must divide by the same gain factor as was used in the compressor and re-insert trailing zeros up to the required wordlength.

A non-uniform quantization process may be used in which the quantizing steps become larger as the signal amplitude increases. As the quantizing steps are made larger, more noise will be suffered, but the noise is arranged to be generated at frequencies where it will not be perceived, or on signal values which occur

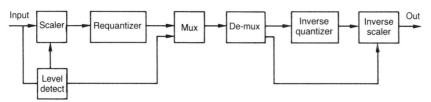

Figure 3.37 Coding gain is obtained by shortening the sample or coefficient wordlength so fewer bits are needed. The input values are scaled or amplified to near maximum amplitude prior to rounding off the low-order bits. The scale factor must be transmitted to allow the process to be reversed at the decoder.

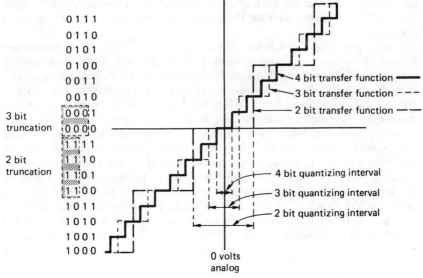

Figure 3.38 Shortening the wordlength of a sample reduces the mumber of codes which can describe the voltage of the waveform. This makes the quantizing steps bigger, hence the term requantizing. It can be seen that simple truncation or omission of the bits does not give analogous behaviour. Rounding is necessary to give the same result as if the larger steps had been used in the original conversion.

relatively infrequently. An audio application of non-uniform quantizing will be found in Section 4.7, and a video application is described in Section 5.11.

Shortening the wordlength of a sample from the LSB end reduces the number of quantizing intervals available without changing the signal amplitude. As Figure 3.38 shows, the quantizing intervals become larger and the original signal is *requantized* with the new interval structure. It will be seen that truncation does not meet the above requirement as it results in signal-dependent offsets because it always rounds in the same direction. Proper numerical rounding is essential for accuracy. Rounding in two's complement is a little more complex than in pure binary. That said, simple truncation is often found in practice.

If the parameter to be requantized is a transform coefficient, the result after decoding is that the frequency which is reproduced has an incorrect amplitude due to quantizing error. If all of the coefficients describing a signal are requantized to roughly the same degree then the error will be uniformly present at all frequencies and will be noise-like.

However, if the data are time-domain samples, as in, for example, a sub-band coder, the result will be different. Although the original audio conversion would have been correctly dithered, the linearizing random element in the low-order bits will be some way below the end of the shortened word. If the word is simply rounded to the nearest integer the linearizing effect of the original dither will be lost and the result will be quantizing distortion. As the distortion takes place in a bandlimited system the harmonics generated will alias back within the band. Where the requantizing process takes place in a sub-band, the distortion products will be confined to that sub-band as shown in Figure 3.39.

Processing for compression 101

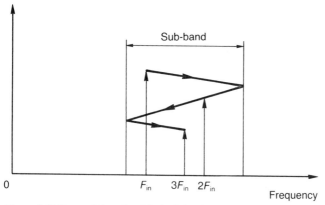

Figure 3.39 Requantizing a band-limited signal causes harmonics which will always alias back within the band.

In practice, the wordlength of samples should be shortened in such a way that the requantizing error is converted to noise rather than distortion. One technique which meets this requirement is to use digital dithering[18] prior to rounding.

Digital dither is a pseudo-random sequence of numbers. If it is required to simulate the analog dither signal of Figure 2.27, then it is obvious that the noise must be bipolar so that it can have an average voltage of zero. Two's complement coding must be used for the dither values.

Figure 3.40 shows a simple digital dithering system for shortening sample wordlength. The output of a two's complement pseudo-random sequence generator of appropriate wordlength is added to input samples prior to rounding. The most significant of the bits to be discarded is examined in order to determine

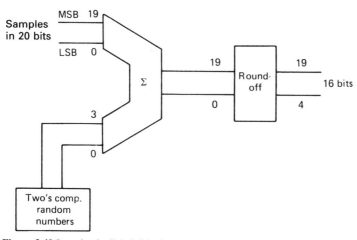

Figure 3.40 In a simple digital dithering system, two's complement values from a random number generator are added to low-order bits of the input. The dithered values are then rounded up or down according to the value of the bits to be removed. The dither linearizes the requantizing.

whether the bits to be removed sum to more or less than half a quantizing interval. The dithered sample is either rounded down, i.e. the unwanted bits are simply discarded, or rounded up, i.e. the unwanted bits are discarded but 1 is added to the value of the new short word. The rounding process is no longer deterministic because of the added dither which provides a linearizing random component.

The probability density of the pseudo-random sequence is important. Lipshitz et al.[19] found that uniform probability density produced noise modulation, in which the amplitude of the random component varies as a function of the amplitude of the samples. A triangular probability density function obtained by adding together two pseudo-random sequences eliminated the noise modulation to yield a signal-independent white-noise component in the least significant bit.

References

1. VAN DEN ENDEN, A.W.M. and VERHOECKX, N.A.M., Digital signal processing: theoretical background. *Philips Tech. Rev.*, **42**, 110–144 (1985)
2. McCLELLAN, J.H., PARKS, T.W. and RABINER, L.R., A computer program for designing optimum FIR linear-phase digital filters. *IEEE Trans. Audio and Electroacoustics* **AU-21**, 506–526 (1973)
3. JAYANT, N.S. and NOLL, P., *Digital Coding of Waveforms: Principles and Applications to Speech and Video.* Prentice-Hall: Englewood Cliffs (1984)
4. THEILE, G., STOLL, G. and LINK, M., Low bit rate coding of high quality audio signals: an introduction to the MASCAM system. *EBU Tech. Review*, No. 230, 158–181 (1988)
5. CHU, P.L., Quadrature mirror filter design for an arbitrary number of equal bandwidth channels. *IEEE Trans. ASSP*, **ASSP-33**, 203–218 (1985)
6. FETTWEIS, A., Wave digital filters: theory and practice. *Proc. IEEE*, **74**, 270–327 (1986)
7. WEISS, P. and CHRISTENSSON, J., Real time implementation of sub-pixel motion estimation for broadcast applications. *IEE Digest*, 1990/128
8. KRANIAUSKAS, P., *Transforms in Signals and Systems*, Chapter 6. Wokingham: Addison Wesley (1992)
9. AHMED, N., NATARAJAN, T. and RAO, K., Discrete cosine transform. *IEEE Trans. Computers*, **C-23**, 90–93 (1974)
10. DE WITH, P.H.N., Data compression techniques for digital video recording. PhD Thesis, Technical University of Delft (1992)
11. GOUPILLAUD, P., GROSSMAN, A. and MORLET, J., Cycle-octave and related transforms in seismic signal analysis. *Geoexploration*, **23**, 85–102. Elsevier (1984/5)
12. DAUBECHIES, I., The wavelet transform, time–frequency localisation and signal analysis. *IEEE Trans. Info. Theory*, **36**, No. 5, 961–1005 (1990)
13. RIOUL, O. and VETTERLI, M., Wavelets and signal processing. *IEEE Signal Process. Mag.*, 14–38 (Oct. 1991)
14. HUFFMAN, J., Wavelets and image compression. Presented at 135th SMPTE Tech. Conf. (Los Angeles 1993), preprint No. 135-98
15. THOMAS, G.A. Television motion measurement for DATV and other applications. *BBC Res. Dept. Rept*, RD 1987/11 (1987)
16. LIMB, J. O. and MURPHY, J.A., Measuring the speed of moving objects from television signals. *IEEE Trans. Commun*, 474–478 (1975)
17. PEARSON, J.J. *et al.*, Video rate image correlation processor. *S.P.I.E.*, Vol. 119 *Application of digital image processing*, IOCC (1977)
18. VANDERKOOY, J. and LIPSHITZ, S.P., Digital dither. Presented at the 81st Audio Engineering Society Convention (Los Angeles, 1986), preprint 2412(C-8)
19. LIPSHITZ, S.P., WANNAMAKER, R.A. and VANDERKOOY, J., Quantization and dither: a theoretical survey. *J. Audio Eng. Soc.*, **40**, 355–375 (1992)

Chapter 4

Audio compression

4.1 Psychoacoustics and masking

By definition, the sound quality of a perceptive coder can only be assessed by human hearing. Equally, a useful perceptive coder can only be designed with a good knowledge of the human hearing mechanism.[1] The acuity of the human ear is astonishing. It can detect tiny amounts of distortion, and will accept an enormous dynamic range. If the ear detects a different degree of impairment between two codecs having the same bit rate in properly conducted tests, we can say that one of them is superior. Thus quality is completely subjective and can only be checked by listening tests. However, any characteristic of a signal which can be heard can also be measured by a suitable instrument. The subjective tests can tell us how sensitive the instrument should be. Then the objective readings from the instrument give an indication of how acceptable a signal is in respect of that characteristic. Instruments for assessing the performance of codecs are currently extremely rare and there remains much work to be done.

The sense we call hearing results from acoustic, mechanical, nervous and mental processes in the ear/brain combination, leading to the term psychoacoustics. It is only possible briefly to introduce the subject here. The interested reader is referred to Moore[2] for an excellent treatment.

Usually, people's ears are at their most sensitive between about 2 kHz and 5 kHz, and although some people can detect 20 kHz at high level, there is much evidence to suggest that most listeners cannot tell if the upper frequency limit of sound is 20 kHz or 16 kHz.[3,4] For a long time it was thought that frequencies below about 40 Hz were unimportant, but it is becoming clear that reproduction of frequencies down to 20 Hz improves reality and ambience.[4,5] The dynamic range of the ear is obtained by a logarithmic response, and certainly exceeds 100 dB. At the extremes of this range, the ear is either straining to hear or is in pain. Neither of these cases can be described as pleasurable or entertaining, and it is hardly necessary to produce recordings of this dynamic range for the consumer since, among other things, he or she is unlikely to have anywhere sufficiently quiet to listen to them. On the other hand extended listening to music whose dynamic range has been excessively compressed is fatiguing.

The sensitivity of the ear to distortion probably deserves more attention than the fidelity of the dynamic range or frequency response. All perceptive audio coding relies on an understanding of the phenomenon of masking, as this determines the audibility of artifacts. The basilar membrane in the ear behaves as a kind of spectrum analyser. The part of the basilar membrane which resonates as a result of an applied sound is a function of the frequency. The basilar membrane is active in that it contains elements which can generate vibration as well as sense it. These are connected in a regenerative fashion so that the Q factor, or frequency selectivity, is higher than it would otherwise be. The high frequencies are detected at the end of the membrane nearest to the eardrum and the low frequencies are detected at the opposite end. The ear analyses with frequency bands, known as critical bands, about 100 Hz wide below 500 Hz and from one-sixth to one-third of an octave wide, proportional to frequency, above this. Critical bands were first described by Fletcher.[6] Later Zwicker experimentally established 24 critical bands,[7] but Moore and Glasberg suggested narrower bands.[8]

In the presence of a complex spectrum, the ear fails to register energy in some bands when there is more energy in a nearby band. The vibration of the membrane in sympathy with a single frequency cannot be localized to an infinitely small area, and nearby areas are forced to vibrate at the same frequency with an amplitude that decreases with distance. Within those areas, other frequencies are excluded unless the amplitude is high enough to dominate the local vibration of the membrane. Thus the Q factor of the membrane is responsible for degree of auditory masking, defined as the decreased audibility of one sound in the presence of another.

The degree of masking depends upon whether the masking tone is a sinusoid, which gives the least masking, or noise.[9] However, harmonic distortion produces widely spaced frequencies, and these are easily detected even in minute quantities by a part of the basilar membrane which is distant from that part which is responding to the fundamental. The masking effect is asymmetrically disposed around the masking frequency.[10] Above the masking frequency, masking is more pronounced, and its extent increases with acoustic level. Below the masking frequency, the extent of masking drops sharply at as much as 90 dB per octave. Clearly very sharp filters are required if noise at frequencies below the masker is to be confined within the masking threshold.

Owing to the resonant nature of the membrane, it cannot start or stop vibrating rapidly. The spectrum sensed changes slowly even if that of the original sound does not. The reduction in information sent to the brain is considerable; masking can take place even when the masking tone begins after and ceases before the masked sound. This is referred to as forward and backward masking.[11] An example of the slowness of the ear is the Haas effect, in which the direction from which a sound is perceived to have come is determined from the first arriving wavefront. Later echoes simply increase the perceived loudness as they have the same spectrum and increase the existing excitation of the membrane.

4.2 Codec level calibration

The functioning of the ear is noticeably level-dependent and perceptive coders take this into account. However, all signal processing takes place in the electrical domain with respect to electrical levels whereas the hearing mechanism operates with respect to sound pressure level. Figure 4.1 shows that in an ideal system the overall gain of the microphones and ADCs is such that the PCM codes have a relationship with sound pressure which is the same as that assumed by the model in the codec. Equally the overall gain of the DAC and loudspeaker system should be such that the sound pressure levels which the codec assumes are those actually

Audio compression 105

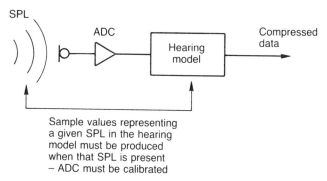

Figure 4.1 Audio coders must be level calibrated so that the psychoacoustic decisions in the coder are based on correct sound pressure levels.

heard. Clearly the gain control of the microphone and the volume control of the reproduction system must be calibrated if the hearing model is to function properly. If, for example, the microphone gain was too low and this was compensated by advancing the loudspeaker gain, the overall gain would be the same but the codec would be fooled into thinking that the sound pressure level was less than it really was and the masking model would not then be appropriate.

The above should come as no surprise as analog audio codecs such as the various Dolby systems have required and implemented line-up procedures and suitable tones.

4.3 Quality measurement

As was seen in Chapter 3, one way in which coding gain is obtained is to requantize sample values to reduce the wordlength. Since the resultant requantizing error results in energy moving from one frequency to another, the masking model is essential to estimate how audible the effect will be. The greater the degree of reduction required, the more precise the model must be. If the masking model is inaccurate, then equipment based upon it may produce audible artifacts under some circumstances. Artifacts may also result if the model is not properly implemented. As a result, development of data reduction units requires careful listening tests with a wide range of source material.[12,13] The presence of artifacts at a given compression factor indicates only that performance is below expectations; it does not distinguish between the implementation and the model. If the implementation is verified, then a more detailed model must be sought. Naturally comparative listening tests are only valid if all the codecs have been level-calibrated.

Properly conducted listening tests are expensive and time-consuming, and alternative methods have been developed which can be used objectively to evaluate the performance of different techniques.[14] The noise-to-masking ratio (NMR) is one such measurement.[15] Figure 4.2 shows how NMR is measured. Input audio signals are fed simultaneously to a data reduction coder and decoder in tandem and to a compensating delay whose length must be adjusted to match the codec delay. At the output of the delay, the coding error is obtained by

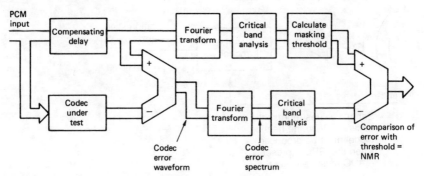

Figure 4.2 The noise-to-masking ratio is derived as shown here.

subtracting the codec output from the original. The original signal is spectrum analysed into critical bands in order to derive the masking threshold of the input audio, and this is compared with the critical band spectrum of the error. The NMR in each critical band is the ratio between the masking threshold and the quantizing error due to the codec. An average NMR for all bands can be computed. A positive NMR in any band indicates that artifacts are potentially audible. Plotting the average NMR against time is a powerful technique, as with an ideal codec the NMR should be stable with different types of programme material. If this is not the case the codec could perform quite differently as a function of the source material. NMR excursions can be correlated with the waveform of the audio input to analyse how the extra noise was caused and to redesign the codec to eliminate it.

Practical systems should have a finite NMR in order to give a degree of protection against difficult signals which have not been anticipated and against the use of post-codec equalization or several tandem codecs which could change the masking threshold. There is a strong argument that professional devices should have a greater NMR than consumer or programme delivery devices.

4.4 The limits

There are, of course, limits to all technologies. Eventually artifacts will be heard as the amount of compression is increased which no amount of detailed modelling will remove. The ear is only able to perceive a certain proportion of the information in a given sound. This could be called the perceptual entropy,[16] and all additional sound is redundant or irrelevant. Data reduction works by removing the redundancy, and clearly an ideal system would remove all of it, leaving only the entropy. Once this has been done, the masking capacity of the ear has been reached and the NMR has reached zero over the whole band. Reducing the data rate further must reduce the entropy, because raising noise further at any frequency will render it audible. In practice the audio bandwidth will have to be reduced in order to keep the noise level acceptable. In MPEG-1 pre-filtering allows data from higher sub-bands to be neglected. MPEG-2 has introduced some low sampling rate options for this purpose. Thus there is a limit to the degree of data reduction which can be achieved even with an ideal coder. Systems which go beyond that limit are not appropriate for high-quality music, but are relevant in news gathering and communications where intelligibility of speech is the criterion.

Interestingly, the data rate out of a coder is virtually independent of the input sampling rate unless the sampling rate is very low. This is because the entropy of the sound is in the waveform, not in the number of samples carrying it.

The compression factor of a coder is only part of the story. All codecs cause delay, and in general the greater the compression the longer the delay. In some applications where the original sound may be heard at the same time as sound which has passed through a codec, a short delay is required.[17] In most applications, the compressed channel will have a constant bit rate, and so a constant compression factor is required. In real programme material, the entropy varies and so the NMR will fluctuate. If greater delay can be accepted, as in a recording application, memory buffering can be used to allow the coder to operate at constant NMR and instantaneously variable data rate. The memory absorbs the instantaneous data rate differences of the coder and allows a constant rate in the channel. A higher effective degree of data reduction will then be obtained.

4.5 Compression applications

One of the fundamental concepts of PCM audio is that the signal-to-noise ratio of the channel can be determined by selecting a suitable wordlength. In conjunction with the sampling rate, the resulting data rate is then determined. In many cases the full data rate can be transmitted or recorded, as is done in DASH, RDAT, CD and the AES/EBU interface. Professional equipment traditionally has operated with higher sound quality than consumer equipment in order to allow for some inevitable quality loss in production, and so there will be less cause to use data reduction. Alternatively, professional equipment might maintain the quality gap by using a smaller degree of data reduction or by employing more sophisticated techniques than consumer devices, although the economics of LSI technology often result in the professional device being forced to use consumer chip sets.

Where there is a practical or economic restriction on channel bandwidth or storage capacity, digital companding becomes essential. Digital audio workstations using hard disks currently cannot compete with tape where large numbers of audio channels are required, but with compression could then do so. In broadcasting, bandwidth is at a premium as sound radio has to share the spectrum with other services. In NICAM TV sound, the new digital sound carrier had to be squeezed into the existing standardized TV channel spacing. NICAM and DAB both use data reduction to conserve data rate. Digital recorders based on RAM or ROM, such as samplers and announcement generators, have no moving parts and are maintenance-free, but suffer from a higher cost per bit than other media. Data reduction reduces the cost per second of audio by allowing longer recording time in a given memory size.

In the Video 8 system, the PCM audio has to share the tape with an analog video signal. Since Video 8 is a consumer product, data reduction is once more used to keep tape consumption moderate.

In DCC it was a goal that the cassette would use conventional oxide tape for low cost, and a simple transport mechanism was a requirement. In order to record on low coercivity tape which is to be used in an uncontrolled consumer environment, wavelengths have to be generous, and track widths need to be relatively wide. In order to allow a reasonably slow tape speed, and hence good playing time, data reduction is necessary.

In the MiniDisc the use of compression allows a very small disk, but this is only one reason for its adoption. MiniDisc is read at the same data rate as CD by the same circuitry. Compression and memory buffering allow the disk to be read discontinuously and if the player is knocked off track it can reposition the pickup whilst continuing to play or record at the other side of the buffer memory.

Broadcasters wishing to communicate over long distances are restricted to the data rate conveniently available in the digital telephone network. Using data reduction, the BBC is able to pass six high-quality audio channels down the available telephone channels of 2048 kbits/s.[18] Before the introduction of stereo, data reduction was used to distribute the sound accompanying television broadcasts by putting two PCM samples inside the line-synchronizing pulses, thus obtaining a sampling rate of twice the TV line rate.[19] The modern dual channel sound-in-syncs (DSIS) systems use a similar approach.

Now that the technology exists to convey television pictures using compressed digital coding, it is obvious that the associated audio should be conveyed in the same way. The MPEG video coding standards accordingly included audio within their scope. The MPEG-1 standard (see Chapter 5) compresses audio and video into about 1.5 Mbits/s. The audio content of MPEG-1 may be used on its own to encode one or two channels at bit rates up to 448 kbits/s. MPEG-2 increases the number of channels to six: Left, Right, Centre, Left surround, Right surround and Subwoofer. In order to retain reverse compatibility with MPEG-1, the MPEG-2 coding converts the six-channel input to a compatible two-channel signal, L_o, R_o, by matrixing.[20] The data from these two channels are encoded in a standard MPEG-1 audio frame, and this is followed in MPEG-2 by an ancillary data frame which an MPEG-1 decoder will ignore. The ancillary frame contains data for any three of the audio channels. An MPEG-2 decoder will extract those three channels in addition to the MPEG-1 frame and then recover all six original channels by an inverse matrix.

4.6 Audio compression techniques

There are many different approaches to audio compression, each allowing a different combination of compression factor and coding delay to be obtained. The simplest systems use either non-uniform requantizing or companding on the whole of the audio band. Non-uniform requantizing can be applied to a uniform quantized PCM signal using a lookup table. With this approach, the range of values in which an input sample finds itself determines the factor by which it will be multiplied. For example, a sample value with the most significant bit reset could be multiplied by 2 to shift the bits up one place. If the two most significant bits are reset, the value could be multiplied by 4 and so on. Constants are then added to allow the range of the compressed sample value to determine the expansion necessary.

A relative of non-uniform requantizing is floating-point coding. Here the sample value is expressed as a mantissa and a binary exponent which determines how the mantissa needs to be shifted to have its correct absolute value on a PCM scale.

Non-uniform requantizing and floating-point coding work on each sample individually. A floating-point system requires one exponent to be carried with each mantissa. An alternative is near-instantaneous companding, or floating-point block coding, where the magnitude of the largest sample in a block is used to determine the value of a common exponent. Sending one exponent per block requires a lower data rate in than true floating point.[21]

Whilst the above systems do reduce the data rate required, only moderate reductions are possible without subjective quality loss. NICAM (Near Instantaneously Companded Audio Multiplex) has an audio data compression factor (output bit rate divided by input bit rate) of 0.7, Video 8 has a factor of 0.8. Applications such as DCC and DAB require a figure of 0.25. In MiniDisc it is 0.2. Compression factors of this kind can only be realized with much more sophisticated techniques. These differ in principle, but share a dependence on complex digital processing which has only recently become economic due to progress in VLSI. These techniques fall into three basic categories: predictive coding, sub-band coding and transform coding.

Predictive coding uses circuitry which uses a knowledge of previous samples to predict the value of the next. It is then only necessary to send the difference between the prediction and the actual value. The receiver contains an identical predictor to which the transmitted difference is added to give the original value.

Sub-band coding splits the audio spectrum up into many different frequency bands to exploit the fact that most bands will contain lower-level signals than the loudest one.

In spectral coding, a Fourier transform of the waveform is computed periodically. Since the transform of an audio signal changes slowly, it need be sent much less often than audio samples. The receiver performs an inverse transform.

Finally, the data may be subject to a lossless binary compression using, for example, a Huffman code.

Practical data-reduction units will usually use some combination of at least two of these techniques along with non-linear or floating-point requantizing of sub-band samples or transform coefficients.

4.7 Non-uniform coding

The digital compression system of Video 8 is illustrated in Figure 4.3. This broadband system reduces samples from 10 bits to 8 bits long and is in addition to an analog compression stage. In order to obtain the most transparent performance, the smallest signals are left unchanged. In a 10 bit system, there are 1024 quantizing intervals, but these are symmetrical about the centre of the range; therefore there are only 512 audio levels in the system, an equal number positive and negative. The 10 bit samples are expressed as signed binary, and only the 9 bit magnitude is compressed to seven bits, following which the sign bit is added on.

The first 16 input levels (0–15) pass unchanged, except that the most significant two bits below the sign bit are removed. The next 48 input levels (16–63) are given a gain of one-half, so that they produce 24 output levels from 16 to 39. The next 256 input levels (64–319) have a gain of one-quarter, so that they produce 64 output levels from 40 to 103. Finally, the remaining 208 levels

110   Audio compression

Input X	Conversion	Output Y
0–15	$Y = X$	0–15
16–63	$Y = \frac{X}{2} + 8$	16–39
64–319	$Y = \frac{X}{4} + 24$	40–103
320–511	$Y = \frac{X}{8} + 44$	104–127

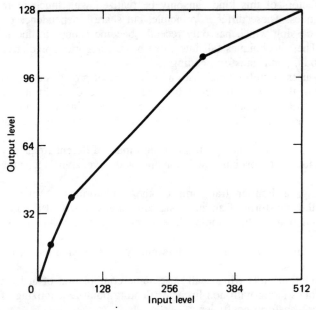

Figure 4.3 The digital compression of PCM samples in 8 mm video. Note that there are only 512 levels in a 10 bit signal, since positive and negative samples can have the same level.

(320–511) have a gain of one-eighth, so they occupy the remaining output levels from 104 to 127. In this way the coarsening of the effective quantizing intervals is matched by the increasing amplitude of the signal, so that the increased noise is masked. The analog compansion further reduces the noise to give remarkably good performance.

4.8 Floating-point coding

In floating-point notation (Figure 4.4), a binary number is represented as a mantissa, which is always a binary fraction with 1 just to the right of the radix point, and an exponent, which is the power of 2 the mantissa has to be multiplied by to obtain the fixed-point number. Clearly the signal-to-noise ratio is now defined by the number of bits in the mantissa, and, as shown in Figure 4.5, this

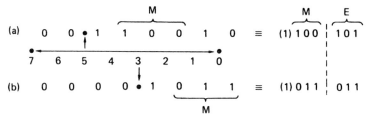

Figure 4.4 In this example of floating-point notation, the radix point can have eight positions determined by the exponent E. The point is placed to the left of the first '1', and the next 4 bits to the right form the mantissa M. As the MSB of the mantissa is always 1, it need not always be stored.

will vary as a sawtooth function of signal level, as the best value, obtained when the mantissa is near overflow, is replaced by the worst value when the mantissa overflows and the exponent is incremented. Floating-point notation is used within DSP chips as it eases the computational problems involved in handling long wordlengths. For example, when multiplying floating-point numbers, only the mantissae need to be multiplied. The exponents are simply added.

Floating-point coding is at its most useful for data reduction when several adjacent samples are assembled into a block so that the largest sample value determines a common exponent for the whole block. This technique is known as floating-point block coding.

In NICAM 728, 14 bit samples are converted to 10 bit mantissae in blocks of 32 samples with a common 3 bit exponent. Figure 4.6 shows that the exponent can have five values, which are spaced 6.02 dB or a factor of 2 in gain apart. When the signal in the block has one or more samples within 6 dB of peak, the

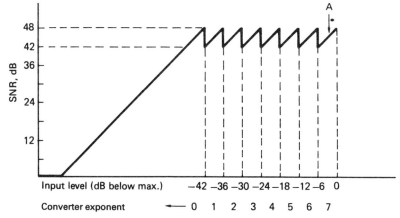

Figure 4.5 In this example of an 8 bit mantissa, 3 bit exponent system, the maximum SNR is 6 dB × 8 = 48 dB with maximum input of 0 dB. As input level falls by 6 dB, the converter noise remains the same, so SNR falls to 42 dB. Further reduction in signal level causes the converter to shift range (point A in the diagram) by increasing the input analog gain by 6 dB. The SNR is restored, and the exponent changes from 7 to 6 in order to cause the same gain change at the receiver. The noise modulation would be audible in this simple system. A longer mantissa word is needed in practice.

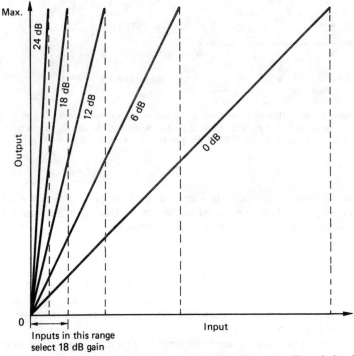

Figure 4.6 The coding scheme of NICAM 728 has five gain ranges. The gain is selected in each block of samples such that the largest coded values are obtained without clipping.

lowest gain is used. If the largest sample in the block has a value more than 6 dB below peak, one high-order bit is inactive and the gain is boosted by a factor of 2 by a 1 bit shift. As a result the system has five transfer functions which are selected to keep the mantissa as large as possible.

The worst case in block coding occurs when there is one large-value sample in an otherwise quiet block. The large sample value causes the system to select a low gain and the quiet part is quantized coarsely, resulting in potential distortion. However, distortion products have to be presented to the ear for some time before they can be detected as a harmonic structure. With a 1 ms block, the distortion is too brief to be heard.

4.9 Predictive coding

Predictive data reduction has many principles in common with sigma-delta modulators. As was seen in Chapter 3, a sigma-delta modulator places a high-order filter around a quantizer so that what is quantized is the difference between the filter output and the actual input. The quantizing error serves to drive the filter in such a way that the difference is cancelled. When the difference is small, the filter has anticipated what the next input voltage will be, which leads it to be called by the alternative title of *predictor*. The same process can be modelled entirely in the digital domain as can be seen in Figure 4.7. The input is

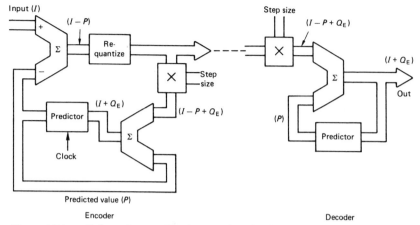

Figure 4.7 A predictive coder resembles the operation of a sigma-delta modulator which has a digital input. See text for details.

conventional PCM, and the predictor output is subtracted from the input by a conventional complement and add process. The difference is then requantized, which requires the numerical value to be divided by the new step size. The remainder is discarded. The requantized difference becomes the data-reduced output. In order to reduce the size of the difference, the requantizing error must be fed back. The requantized difference must be returned to its original numerical value (except for the requantizing error) by multiplying it by the same step size that was used in the requantizer. This is the function of the inverse quantizer. The inverse quantized difference is then added to the predictor output. The next input to the predictor is thus the current input plus the requantizing error.

The decoder contains the same predictor and inverse quantizer as the encoder. The inverse quantizer returns the differences to their correct numerical value. The predictor is driven from the system output and will make the same prediction as the encoder. The difference is added to the predictor output to re-create the original value of the sample, which also becomes the next predictor input.

Further data reduction can be obtained by making the requantizing steps of variable size. When the difference signal is large, the step size increases to prevent saturation. When the difference signal is small, the step size reduces to increase resolution. In order to avoid transmitting the step size along with the difference data, it is possible to make the requantizer adaptive. In this case the step size is based on the magnitude of recent requantizer outputs. As the output increases towards the largest code value, the step size increases to avoid clipping. Figure 4.8 shows how an adaptive quantizer can be included in the predictive coder. Clearly the same step size needs to be input both to the inverse quantizer in the coder and to the same device in the decoder. Clearly if the step size is obtained from the coder output, the decoder can also re-create it locally so that the same step size is used throughout.

Predictive coders work by finding the redundancy in an audio signal. For example, the fundamental of a note from a musical instrument will be a sine wave which changes in amplitude quite slowly. The predictor will learn the pitch quite quickly and remove it from the data stream which then contains the

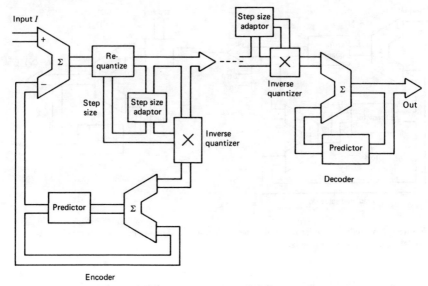

Figure 4.8 A predictive coder incorporating an adaptive quantizer bases the step size on recent outputs. The decoder can re-create the same step size locally so it need not be sent.

harmonics only. These are at a lower level than the fundamental and require less data to represent them. When a transient arrives, the predictor will be unprepared, and a large difference will be computed. The adaptive quantizer will react by increasing the step size to prevent clipping.

DPCM coders have the advantage that they work on the signal waveform in the time domain and need a relatively short signal history to operate. Thus they cause a relatively short delay in the coding and decoding stages. A further advantage is that the differential data are actually less sensitive to bit errors than PCM. This is because the difference signals represent a small part of the final signal amplitude.

4.10 Sub-band coding

Sub-band data reduction takes advantage of the fact that real sounds do not have uniform spectral energy. The wordlength of PCM audio is based on the dynamic range required and this is generally constant with frequency, although any pre-emphasis will affect the situation. When a signal with an uneven spectrum is conveyed by PCM, the whole dynamic range is occupied only by the loudest spectral component, and all of the other components are coded with excessive headroom. In its simplest form, sub-band coding[22] works by splitting the audio signal into a number of frequency bands and companding each band according to its own level. Bands in which there is little energy result in small amplitudes which can be transmitted with short wordlength. Thus each band results in variable-length samples, but the sum of all the sample wordlengths is less than that of PCM and so a coding gain can be obtained. Sub-band coding is not restricted to the digital domain; the analog Dolby noise reduction systems use it extensively.

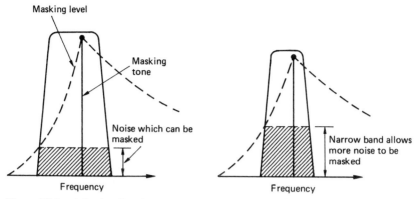

Figure 4.9 In sub-band coding the worst case occurs when the masking tone is at the top edge of the sub-band. The narrower the band, the higher the noise level which can be masked.

The number of sub-bands to be used depends upon what other reduction technique is to be combined with the sub-band coding. If it is intended to use reduction based on auditory masking, the sub-bands should preferably be narrower than the critical bands of the ear, and therefore a large number will be required; ISO/MPEG and PASC, for example, use 32 sub-bands. Figure 4.9 shows the critical condition where the masking tone is at the top edge of the sub-band. It will be seen that the narrower the sub-band, the higher the requantizing noise that can be masked. The use of an excessive number of sub-bands will, however, raise complexity and the coding delay, as well as risking pre-echo on transients exceeding the temporal masking.

The QMF technique is restricted to bands of equal width. It might be thought that this is a drawback because the critical bands of the ear are non-uniform. In fact this is only a problem when very low bit rates are required. In all cases it is the masking model of hearing which must have correct critical bands. This model can then be superimposed on bands of any width to determine how much masking and therefore coding gain is possible. Uniform width sub-bands will not be able to obtain as much masking as bands which are matched to critical bands, but for many applications the additional coding gain is not worth the added filter complexity.

On the other hand, if used in conjunction with predictive coding, relatively few bands are required. The apt-x100 system, for example, uses only four bands as simulations showed that a greater number gave diminishing returns.[23]

4.11 Transform coding

Audio is usually considered to be a time-domain waveform as this is what emerges from a microphone. As has been seen in Chapter 3, Fourier analysis allows any periodic waveform to be represented by a set of harmonically related components of suitable amplitude and phase. In theory it is perfectly possible to decompose an input waveform into its constituent frequencies and phases, and to record or transmit the transform. The transform can then be reversed and the original waveform will be precisely re-created. Although one can think of exceptions, the transform of a typical audio waveform changes relatively slowly.

The slow speech of an organ pipe or a violin string, or the slow decay of most musical sounds, allow the rate at which the transform is sampled to be reduced, and a coding gain results. A further coding gain will be achieved if the components which will experience masking are quantized more coarsely.

In practice there are some difficulties: real sounds are not periodic, but contain transients which transformation cannot accurately locate in time. The solution to this difficulty is to cut the waveform into short segments and then to transform each individually. The delay is reduced, as is the computational task, but there is a possibility of artifacts arising because of the truncation of the waveform into rectangular time windows. A solution is to use window functions (see Chapter 3) and to overlap the segments as shown in Figure 4.10. Thus every input sample appears in just two transforms, but with variable weighting depending upon its position along the time axis. Although it appears from the diagram that twice as much data will be generated, in fact certain transforms can eliminate the redundancy.

Figure 4.10 Transform coding can only be practically performed on short blocks. These are overlapped using window functions in order to handle continuous waveforms.

The DFT (discrete frequency transform) does not produce a continuous spectrum, but instead produces coefficients at discrete frequencies. The frequency resolution (i.e. the number of different frequency coefficients) is equal to the number of samples in the window. If overlapped windows are used, twice as many coefficients are produced as are theoretically necessary. In addition the DFT requires intensive computation, owing to the requirement to use complex arithmetic to render the phase of the components as well as the amplitude. An alternative is to use discrete cosine transforms (DCT). These are advantageous when used with overlapping windows. In the modified discrete cosine transform (MDCT),[24] windows with 50 per cent overlap are used. Thus twice as many coefficients as necessary are produced. These are subsampled by a factor of two to give a critically sampled transform, which results in potential aliasing in the frequency domain. However, by making a slight change to the transform, the alias products in the second half of a given window are equal in size but of opposite polarity to the alias products in the first half of the next window, and so will be cancelled on reconstruction. This is the principle of time domain aliasing cancellation (TDAC).

The requantizing in the coder raises the quantizing noise in the frequency bin, but it does so over the entire duration of the block. Figure 4.11 shows that if a transient occurs towards the end of a block, the decoder will reproduce the waveform correctly, but the quantizing noise will start at the beginning of the block and may result in a pre-echo where the noise is audible before the transient.

The solution is to use a variable time window according to the transient content of the audio waveform. When musical transients occur, short blocks are necessary and the frequency resolution and hence the coding gain will be low. At

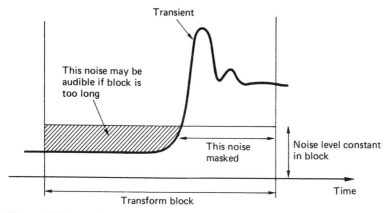

Figure 4.11 If a transient occurs towards the end of a transform block, the quantizing noise will still be present at the beginning of the block and may result in a pre-echo where the noise is audible before the transient.

other times the blocks become longer and the frequency resolution of the transform rises, allowing a greater coding gain.

The transform of an audio signal is computed in the main signal path in a transform coder, and has sufficient frequency resolution to drive the masking model directly. However, in certain sub-band coders the frequency resolution of the filter bank is good enough to offer a high coding gain, but not good enough to drive the masking model accurately, particularly in respect of the steep slope on the low-frequency side of the masker. In order to overcome this problem, a transform will often be computed for control purposes in a side chain rather than in the main audio path, and so the accuracy in respects other than frequency resolution need not be so high. This approach also permits the use of equal width sub-bands in the main path.

4.12 A simple sub-band coder

Figure 4.12 shows the block diagram of a simple sub-band coder. At the input, the frequency range is split into sub-bands by a filter bank such as a quadrature mirror filter. The output data rate of the filter bank is no higher than the input rate because each band has been heterodyned to a frequency range from DC upwards. The decomposed sub-band data are then assembled into blocks of fixed size, prior to reduction. Whilst all sub-bands may use blocks of the same length, some coders may use blocks which get longer as the sub-band frequency becomes lower. Sub-band blocks are also referred to as frequency bins.

The coding gain is obtained as the waveform in each band passes through a requantizer. The requantization is achieved by multiplying the sample values by a constant and rounding up or down to the required wordlength. For example, if in a given sub-band the waveform is 36 dB down on full scale, there will be at least six bits in each sample which merely replicate the sign bit. Multiplying by 2^6 will bring the high-order bits of the sample into use, allowing bits to be lost at the lower end by rounding to a shorter wordlength. The shorter the wordlength, the greater the coding gain, but the coarser the quantization steps and therefore the level of quantization error.

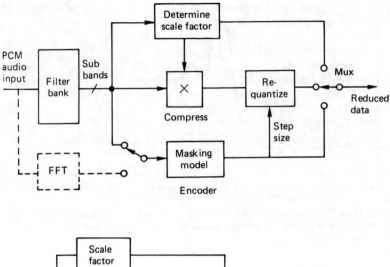

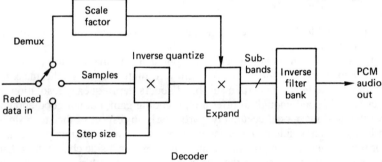

Figure 4.12 A simple sub-band coder. The bit allocation may come from analysis of the sub-band energy, or, for greater reduction, from a spectral analysis in a side chain.

If a fixed data reduction factor is employed, the size of the coded output block will be fixed. The requantization wordlengths will have to be such that the sum of the bits from each sub-band equals the size of the coded block. Thus some sub-bands can have long wordlength coding if others have short wordlength coding. The process of determining the requantization step size, and hence the wordlength in each sub-band, is known as bit allocation. The bit allocation may be performed by analysing the power in each sub-band, or by a side chain which performs a spectral analysis or transform of the audio. The complexity of the bit allocation depends upon the degree of compression required. The spectral content is compared with an auditory masking model to determine the degree of masking which is taking place in certain bands as a result of higher levels in other bands. Where masking takes place, the signal is quantized more coarsely until the quantizing noise is raised to just below the masking level. The coarse quantization requires shorter wordlengths and allows a coding gain. The bit allocation may be iterative as adjustments are made to obtain the best NMR within the allowable data rate.

The samples of differing wordlength in each bin are then assembled into the output coded block. Unlike a PCM block, which contains samples of fixed wordlength, a coded block contains many different wordlengths and these can vary from one block to the next. In order to deserialize the block into samples of various wordlength and demultiplex the samples into the appropriate frequency bins, the decoder has to be told what bit allocations were used when it was packed, and some synchronizing means is needed to allow the beginning of the block to be identified. Demultiplexing can be done by including the transform of the block which was used to determine the allocation. If the decoder has the same allocation model, it can determine what the coder must have done from the transform, and demultiplex the block accordingly. Once all of the samples are back in their respective frequency bins, the level of each bin is returned to the original value. This is achieved by reversing the gain increase which was applied before the requantizer in the coder. The degree of gain reduction to use in each bin also comes from the transform. The sub-bands can then be recombined into a continuous audio spectrum in the output filter which produces conventional PCM of the original wordlength.

The degree of compression is determined by the bit allocation system. It is not difficult to change the output block size parameter to obtain a different compression. The bit allocator simply iterates until the new block size is filled. Similarly the decoder need only deserialize the larger block correctly into coded samples and then the expansion process is identical except for the fact that expanded words contain less noise. Thus codecs with varying degrees of compression are available which can perform different bandwidth/performance tasks with the same hardware.

4.13 Data reduction formats

There are currently a large number of data reduction formats available either as commercially developed products or as international standards under a variety of names. This section sets out the relationships between the major formats.

The ISO (International Standards Organization) and the IEC (International Electrotechnical Commission) recognized that data reduction would have an important part to play in future digital video products and in 1988 established the ISO/IEC/MPEG (Moving Picture Experts Group) to compare and assess various coding schemes in order to arrive at an international standard. The terms of reference were extended the same year to include audio and the MPEG/Audio group was formed.

As part of the Eureka 147 project, a system known as MUSICAM[25] (Masking pattern adapted Universal Sub-band Integrated Coding And Multiplexing) was developed jointly by CCETT in France, IRT in Germany and Philips in the Netherlands. MUSICAM was designed to be suitable for DAB.

As a parallel development, the ASPEC[26] (Adaptive Spectral Perceptual Entropy Coding) system was developed from a number of earlier systems as a joint proposal by AT&T Bell Labs, Thomson, the Fraunhofer Society and CNET. ASPEC was designed for high degrees of compression to allow audio transmission on ISDN.

These two systems were both fully implemented by July 1990 when comprehensive subjective testing took place at the Swedish Broadcasting Corporation.[12,27,28] As a result of these tests, the MPEG/Audio group combined the attributes of both ASPEC and MUSICAM into a draft standard[29] having three levels of complexity and performance.

As has been seen above, coders can be operated at various compression factors. The performance of a given codec will be improved if the degree of compression is reduced, and so for moderate compression, a simple codec will be more cost-effective. On the other hand, as the degree of compression is increased, the codec performance will be reduced, and it will be necessary to switch to a higher layer, with attendant complexity, to maintain quality. ISO/MPEG coding allows input sampling rates of 32, 44.1 and 48 kHz and supports output bit rates of 32, 48, 56, 64, 96, 112, 128, 192, 256 and 384 kbits/s. The transmission can be mono, dual channel (e.g. bilingual), stereo and joint stereo which is where advantage is taken of redundancy between the two audio channels.

ISO Layer 1 is a simplified version of MUSICAM which is appropriate for the mild compression applications at low cost. It is very similar to PASC. ISO Layer II is identical to MUSICAM and is very likely to be used for DAB. ISO Layer III is a combination of the best features of ASPEC and MUSICAM and is mainly applicable to telecommunications.

4.14 ISO Layer I – simplified MUSICAM

The simplified MUSICAM system is a sub-band code based on the block diagram of Figure 4.12. A polyphase quadrature mirror filter network divides the audio spectrum into 32 equally spaced bands. Constant-size blocks are used, containing 12 samples in each sub-band. The block size was based on the pre-masking phenomenon of Figure 4.11. Like NICAM, the samples in each sub-band block are block-compressed according to the peak value in the block. A 6 bit scale factor is used for each sub-band. The sub-bands themselves are used as a spectral analysis of the input in order to determine the bit allocation. The mantissae of the samples in each block are requantized according to the bit allocation. The bit allocation data are multiplexed with the scale factors and the requantized sub-band samples to form the coded message frames. The bit allocation codes are necessary to allow the decoder correctly to assemble the variable-length samples.

4.15 ISO Layer II – MUSICAM

ISO Layer II is identical to MUSICAM. The same filterbank as Layer I is used, and the blocks are the same size. The same block-companding scheme is used. In order to give better spectral resolution than the filterbank, a side chain FFT is computed having 1024 points, resulting in an analysis of the audio spectrum eight times better than the sub-band width. The FFT drives the masking model which controls the bit allocation.

Whilst the block-companding scheme is the same as in layer I, not all of the scale factors are transmitted, because they contain a degree of redundancy on real programme material. The scale factor of successive blocks in the same band exceeds 2 dB less than 10 per cent of the time, and advantage is taken of this characteristic by analysing sets of three successive scale factors. On stationary programme, only one scale factor out of three is sent. As transient content increases in a given sub-band, two or three scale factors will be sent. A scale factor select code is also sent to allow the decoder to determine what has been sent in each sub-band. This technique effectively halves the scale factor bit rate.[30]

The requantized samples in each sub-band, bit allocation data, scale factors and scale factor select codes are multiplexed into the output bit stream.

4.16 ISO Layer III

This is the most complex layer of the ISO standard, and is only really necessary when the most severe data rate constraints must be met with high quality. It is a transform code based on the ASPEC system with certain modifications to give a degree of commonality with Layer II. The original ASPEC coder used a direct MDCT on the input samples. In Layer III this was modified to use a hybrid transform incorporating the existing polyphase 32 band QMF of Layers I and II. In Layer III, the 32 sub-bands from the QMF are each processed by a 12 band MDCT to obtain 384 output coefficients. Two window sizes are used to avoid pre-echo on transients. The window switching is performed by the psycho-acoustic model. It has been found that pre-echo is associated with the entropy in the audio rising above the average value.

A highly accurate perceptive model is used to take advantage of the high frequency resolution available. Non-uniform quantizing is used, along with Huffman coding. This is a technique where the most common code values are allocated the shortest wordlength.

4.17 apt-x100

The apt-x100 codec uses the predictive coding techniques described in Section 4.12 in four sub-bands to achieve compression to 0.25 of the original bit rate. The sub-bands are derived with quadrature mirror filters, but in each sub-band a continuous predictive coding takes place which is matched by a continuous decoding at the receiver. Blocks are not used for coding, but only for packing the difference values for transmission. The output block consists of 2048 bits and commences with a synchronizing pattern which enables the decoder correctly to assemble difference values and attribute them to the appropriate sub-band. The decoder must see three sync patterns at the correct spacing before locking is considered to have occurred. The synchronizing system is designed so that four compressed data streams can be compressed into one 16 bit channel and correctly demultiplexed at the decoders.

With a continuous DPCM coder there is no reliance on temporal masking, but adaptive coders which vary the requantizing step size will need to have a rapid step size attack in order to avoid clipping on transients. Following the transient, the signal will often decay more quickly than the step size, resulting in excessively coarse requantization. During this period, temporal masking prevents audibility of the noise. As the process is waveform-based rather than spectrum-based, neither an accurate model of auditory masking nor a large number of sub-bands are necessary. As a result, apt-x100 can operate over a wide range of sampling rates without adjustment whereas in the majority of coders changing the sampling rate means that the sub-bands have different frequencies and will require different masking parameters. A further salient advantage of the predictive approach is that the delay through the codec is less than 4 ms, which is advantageous for live (rather than recorded) applications.

4.18 Dolby AC-2

The Dolby AC-2 is in fact a family of transform coders based on TDAC which allow various compromises between coding delay and bit rate to be used.[10] Figure 4.13 shows the generic block diagram of the AC-2 coder. Input audio is passed through a 50 per cent overlapped critically sampled TDAC transform which uses alternate modified sine and cosine transforms. The high-frequency resolution coefficients are selectively combined in sub-bands which approximate the critical bands. Coefficients in each sub-band are normalized

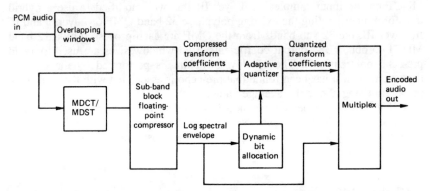

Figure 4.13 Block diagram of the Dolby AC-2 coder. See text for details.

and expressed in floating-point block notation with common exponent. The exponents in fact represent the logarithmic spectral envelope of the signal and can be used to drive the perceptive model which operates the bit allocation. The mantissae of the transform coefficients are then requantized according to the bit allocation. The bit allocation technique is conservative, as it is primarily fixed with only around 20 per cent of the bits being adaptively allocated. The data stream consists of the requantized coefficients and the log spectral envelope in the shape of the exponents. The receiver uses the log spectral envelope to deserialize the coefficients into the correct wordlengths prior to returning them to fixed-point notation. Inverse transforms are then computed, followed by a weighted overlapping of the windows to obtain PCM data.

4.19 PASC

Precision Adaptive Sub-band Coding (PASC) was developed by Philips for the Digital Compact Cassette (DCC)[31] and is similar to the ISO/MPEG Layer I system with which Philips were also involved. The DCC transport operates at a data rate of 384 kbits/s and PASC is designed to give high-fidelity stereo performance at that rate, corresponding to a reduction factor of 0.25.

PASC uses 32 equal sub-bands which are produced with a polyphase QMF having 512 taps. Sampling rates of 32, 44.1 and 48 kHz can be used, and this results in the sub-bands having widths of 500, 689 and 750 Hz and coding delays

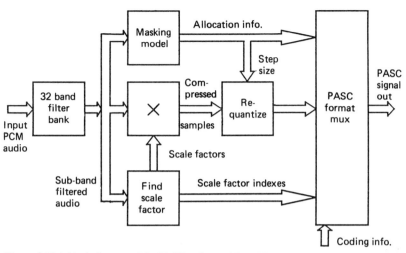

Figure 4.14 A block diagram of the PASC coder used in DCC. This is similar to the ISO/MPEG Layer I system.

of 15, 12 and 10 ms respectively. In each sub-band, samples are assembled into blocks of twelve. A block will have a period of 8 ms at 48 kHz.

Figure 4.14 shows a block diagram of the PASC coder. Inputs of up to 18 bit PCM are sub-band filtered, resulting in 24 bit sub-band samples. The signals in each band are floating-point block coded. A single scale factor index is determined from the largest sample in the block. Blocks containing small numerical values are multiplied until the MSB of the largest value is active. The multiplier is the exponent or scale factor which is passed onto the formatter for transmission. The maximum length of the mantissa is 15 bits, giving a resolution of 92 dB, but this resolution can be placed anywhere over a range of +6 to −118 dB in 2 dB steps by the 6 bit scale factor index.

As the degree of compression is moderate, a sufficiently accurate input spectrum can be obtained by calculating the signal power in each sub-band. This is compared with an absolute masking threshold to establish the in-band masking level where signal components are above absolute masking. Levels in bands containing signals above the absolute masking threshold are then used to compute the degree of out-of-band masking available in an individual band due to signals in other bands. The outcome of this process is that in each sub-band the resolution necessary to prevent noise being audible is established. From this the wordlength of the mantissae in each block can be determined. As the sub-band samples were normalized by the scale factor, requantizing to the chosen wordlength is easy. The bit allocation must also be conveyed to the decoder. The various contributions to the coded data are combined into one data stream by the PASC formatter.

Figure 4.15 shows the format of the PASC data stream. The frame begins with a sync pattern to reset the phase of deserialization, and a header which describes the sampling rate and any use of pre-emphasis. Following this is a block of 32 four-bit allocation codes. These specify the wordlength used in each sub-band

124 Audio compression

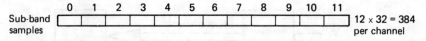

From these are calculated:

    32 allocation information units, one for each sub-band
    32 x 4 bits = 128 bits

    32 scale factor indices, one for each sub-band transferred
    32 x 6 bits = 192 bits max.

    32 x 12 PASC coded sub-band samples
    0 ... 15 bits/sample

Synchronization pattern and coding info. (sample freq., emphasis) are added

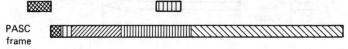

Figure 4.15 The PASC data frame showing the allocation codes, the scale factors and the sub-band samples.

and allow the PASC decoder to deserialize the sub-band sample block. This is followed by a block of 32 six-bit scale factor indices, which specify the gain given to each band during normalization. The last block contains 32 sets of 12 samples. These samples vary in wordlength from one block to the next, and can be from 0 to 15 bits long. The PASC deserializer has to use the 32 allocation information codes to work out how to deserialize the sample block into individual samples of variable length.

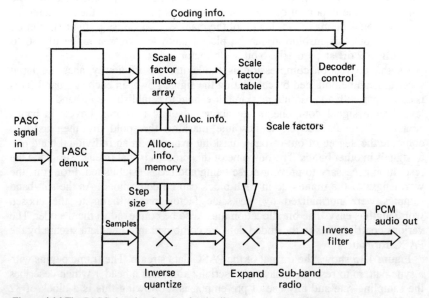

Figure 4.16 The PASC decoder. See text for details.

Whatever the sampling rate used, the PASC frame is always 384 sample periods long. The tape frame rate is constant and so between 14 and 21 PASC frames are accommodated in a tape frame according to the sampling rate. The PASC frames must be smaller at higher sampling rates, and the bit allocation is controlled accordingly. At 48 kHz, the average length of the samples is just under 4 bits, whereas at 32 kHz it is just under 6 bits. Thus in DCC the tape speed is independent of the sampling rate.

The PASC decoder is shown in Figure 4.16. The PASC frame is deserialized using the sync pattern and the variable-length samples are assembled using the allocation codes. The variable-length samples are returned to 15 bit wordlength by adding zeros. The scale factor indices are then used to determine multiplication factors used to return the waveform in each sub-band to its original level.

The 32 sub-band signals are then merged into one spectrum by the synthesis filter which returns every sub-band to the correct place in the audio spectrum.

4.20 The ATRAC coder

The ATRAC (Adaptive TRansform Acoustic Coder) coder was developed by Sony and is used in MiniDisc. ATRAC uses a combination of sub-band coding and modified discrete cosine transform (MDCT) coding. Figure 4.17 shows a block diagram of an ATRAC coder. The input is 16 bit PCM audio. This passes through a quadrature mirror filter which splits the audio band into two halves. The lower half of the spectrum is split in half once more, and the upper half passes through a compensating delay. Each frequency band is formed into blocks, and each block is then subject to a modified discrete cosine transform. The frequencies of the DCT are grouped into a total of 52 frequency bins which are of varying bandwidth according to the width of the critical bands in the hearing mechanism. The coefficients in each frequency bin are then companded and requantized as for the sub-band coder in Section 4.12. The requantizing once more is performed on a bit allocation basis using a masking model.

In order to prevent pre-echo, ATRAC selects blocks as short as 1.45 ms in the case of large transients, but the block length can increase in steps up to a maximum of 11.6 ms when the waveform has stationary characteristics. The block size is selected independently in each of the three bands.

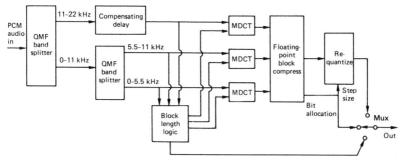

Figure 4.17 The ATRAC coder uses variable-length blocks and MDCT in three sub-bands.

The coded data includes side chain parameters which specify the block size and the wordlength of the coefficients in each frequency bin.

Decoding is straightforward. The bit stream is deserialized into coefficients of various wordlengths and block durations according to the side chain data. The coefficients are then used to control inverse DCTs which re-create time-domain waveforms in the three sub-bands. These are recombined in the output filter to produce the conventional PCM output. In MiniDisc, the ATRAC coder compresses 44.1 kHz 16 bit PCM to 0.2 of the original data rate.

References

1. JOHNSTON, J.D., Transform coding of audio signals using perceptual noise criteria. *IEEE J. Selected Areas in Comms.*, **JSAC-6**, 314–323 (1988)
2. MOORE, B.C.J., *An Introduction to the Psychology of Hearing*. London: Academic Press (1989)
3. MURAOKA, T., IWAHARA, M. and YAMADA, Y., Examination of audio bandwidth requirements for optimum sound signal transmission. *J. Audio Eng. Soc.*, **29**, 2–9 (1982)
4. MURAOKA, T., YAMADA, Y. and YAMAZAKI, M., Sampling frequency considerations in digital audio. *J. Audio Eng. Soc.*, **26**, 252–256 (1978)
5. FINCHAM, L.R., The subjective importance of uniform group delay at low frequencies. Presented at the 74th Audio Engineering Society Convention (New York, 1983), preprint 2056(H-1)
6. FLETCHER, H., Auditory patterns. *Rev. Modern Physics*, **12**, 47–65 (1940)
7. ZWICKER, E., Subdivision of the audible frequency range into critical bands. *J. Acoust. Soc. Amer.*, **33**, 248 (1961)
8. MOORE, R. and GLASBERG, B., Formulae describing frequency selectivity as a function of frequency and level, and their use in calculating excitation patterns. *Hearing Research*, **28**, 209–225 (1987)
9. EHMER, R.H., Masking of tones vs. noise bands. *J. Audio Eng. Soc.*, **31**, 1253–1256 (1959)
10. FIELDER, L.D. and DAVIDSON, G.A., AC-2: a family of low complexity transform based music coders. *Proc. 10th. Int. Audio Eng. Soc. Conf.*, 57–70, New York: Audio Eng. Soc. (1991)
11. CARTERETTE, E.C. and FRIEDMAN, M.P., *Handbook of Perception*, 305–319. New York: Academic Press (1978)
12. GREWIN, C. and RYDEN, T., Subjective assessments on low bit-rate audio codecs. *Proc. 10th. Int. Audio Eng. Soc. Conf.*, 91–102. New York: Audio Eng. Soc. (1991)
13. GILCHRIST, N.H.C., Digital Sound: the selection of critical programme material and preparation of the recordings for CCIR tests on low bit rate codecs. *BBC Res. Dept. Rep.*, RD 1993/1
14. COLOMES, C. and FAUCON, G., A perceptual objective measurement system (POM) for the quality assessment of perceptual codecs. Presented at 96th Audio Eng. Soc. Conv. Amsterdam (1994), preprint No. 3801 (P4.2)
15. BRANDENBURG, K. and SEITZER, D., Low bit rate coding of high quality digital audio: algorithms and evaluation of quality. *Proc. 7th Int. Audio Eng. Soc. Conf.*, 201–209. New York: Audio Eng. Soc. (1989)
16. JOHNSTON, J., Estimation of perceptual entropy using noise masking criteria. *ICASSP*, 2524–2527 (1988)
17. GILCHRIST, N.H.C., Delay in broadcasting operations. Presented at 90th Audio Eng. Soc. Conv. (1991), preprint 3033
18. McNALLY, G.W., Digital audio in broadcasting. *IEEE ASSP Magazine*, **2**, 26–44 (1985)
19. JONES, A.H., A PCM sound-in-syncs distribution system. General description. *BBC Res. Dept Rept*, 1969/35
20. BONICEL, P. et al., A real time ISO/MPEG2 Multichannel decoder. Presented at 96th Audio Eng. Soc. Conv. (1994), preprint No. 3798 (P3.7)
21. CAINE, C.R., ENGLISH, A.R. and O'CLAREY, J.W.H., NICAM-3: near-instantaneous companded digital transmission for high-quality sound programmes. *J. IERE*, **50**, 519–530 (1980)
22. CROCHIERE, R.E., Sub-band coding. *Bell System Tech. J.*, **60**, 1633–1653 (1981)
23. SMYTH, S.M.F. and McCANNY, J.V., 4-bit hi-fi: high quality music coding for ISDN and broadcasting applications. *Proc. ASSP*, 2532–2535 (1988)
24. PRINCEN, J.P., JOHNSON, A. and BRADLEY, A.B., Sub-band/transform coding using filter bank designs based on time domain aliasing cancellation. *Proc. ICASSP*, 2161–2164 (1987)

25. WIESE, D., MUSICAM: flexible bitrate reduction standard for high quality audio. Presented at Digital Audio Broadcasting Conference (London, March 1992)
26. BRANDENBURG, K., ASPEC coding. *Proc. 10th. Audio Eng. Soc. Int. Conf.*, 81–90. New York: Audio Eng. Soc. (1991)
27. ISO/IEC JTC1/SC2/WG11 N0030: MPEG/AUDIO test report. Stockholm (1990)
28. ISO/IEC JTC1/SC2/WG11 MPEG 91/010 The SR report on: The MPEG/AUDIO subjective listening test. Stockholm (1991)
29. ISO/IEC JTC1/SC2/WG11 Committee draft 11172
30. WIESE, D., MUSICAM: flexible bitrate reduction standard for high quality audio. Presented at Digital Audio Broadcasting Conference (London, March 1992)
31. LOKHOFF, G.C.P., DCC: Digital compact cassette. *IEEE Trans. Consum. Electron.*, **CE-37**, 702–706 (1991)

Chapter 5
Video compression

In this chapter the principles of video compression are explored, leading to descriptions of the major compression standards. As some compression standards do not support interlace and some do, this chapter uses the term 'picture' to mean an image of any kind at one point on the time axis. This could be a field in interlaced systems and a frame in non-interlaced systems. The terms field and frame will be used only when the distinction is important.

5.1 The eye

All television signals ultimately excite some response in the eye and the viewer can only describe the result subjectively. Familiarity with the functioning and limitations of the eye is essential to an understanding of video compression.

The simple representation of Figure 5.1 shows that the eyeball is nearly spherical and is swivelled by muscles. The space between the cornea and the lens is filled with transparent fluid known as *aqueous humour*. The remainder of the eyeball is filled with a transparent jelly known as *vitreous humour*. Light enters the cornea, and the amount of light admitted is controlled by the pupil in the iris. Light entering is involuntarily focused on the retina by the lens in a process called *visual accommodation*. The lens is the only part of the eye which is not nourished by the bloodstream and its centre is technically dead. In a young person the lens is flexible and muscles distort it to perform the focusing action. In old age the lens loses some flexibility and causes *presbyopia* or limited accommodation. In some people the length of the eyeball is incorrect resulting in *myopia* (short-sightedness) or *hypermetropia* (long-sightedness). The cornea should have the same curvature in all meridia, and if this is not the case, *astigmatism* results.

The retina is responsible for light sensing and contains a number of layers. The surface of the retina is covered with arteries, veins and nerve fibres and light has to penetrate these in order to reach the sensitive layer. This contains two types of discrete receptors known as *rods* and *cones* from their shape. The distribution and characteristics of these two receptors are quite different. Rods dominate the periphery of the retina whereas cones dominate a central area known as the *fovea* outside which their density drops off. Vision using the rods is monochromatic and has poor resolution but remains effective at very low light levels, whereas the cones provide high resolution and colour vision but require more light. Figure 5.2 shows how the sensitivity of the retina slowly increases in response to entering

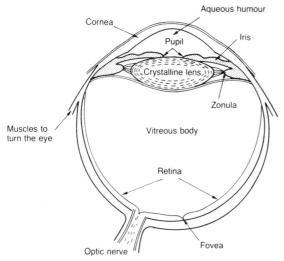

Figure 5.1 The eyeball is in effect a living camera except for the lens which receives no blood supply and is technically dead.

darkness. The first part of the curve is the adaptation of cone or *photopic* vision. This is followed by the greater adaptation of the rods in *scotopic* vision. At such low light levels the fovea is essentially blind and small objects which can be seen in the peripheral rod vision disappear when stared at.

The cones in the fovea are densely packed and directly connected to the nervous system, allowing the highest resolution. Resolution then falls off away from the fovea. As a result the eye must move to scan large areas of detail. The

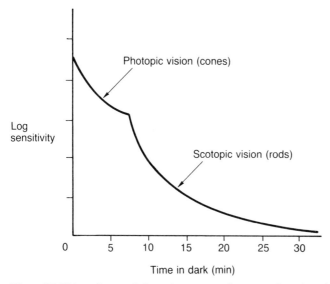

Figure 5.2 Vision adapts to darkness in two stages known as photopic and scotopic vision.

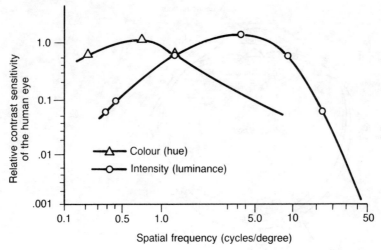

Figure 5.3 The response of the eye to static detail falls of at both low and high spatial frequencies.

image perceived is not just a function of the retinal response, but is also affected by processing of the nerve signals. The overall acuity of the eye can be displayed as a graph of the response plotted against the degree of detail being viewed. Detail is generally measured in lines per millimetre or cycles per picture height, but this takes no account of the distance from the eye. A better unit for eye resolution is one based upon the subtended angle of detail as this will be independent of distance. Units of cycles per degree are then appropriate. Figure 5.3 shows the response of the eye to static detail. Note that the response to very low frequencies is also attenuated. An extension of this characteristic allows the vision system to ignore the fixed pattern of shadow on the retina due to the nerves and arteries.

The retina does not respond instantly to light, but requires between 0.15 and 0.3 s before the brain perceives an image. Scotopic vision experiences a greater delay than photopic vision as more processes are required. Images are retained for about 0.1 s – the phenomenon of *persistence of vision*. Flashing lights are perceived to flicker until the *critical flicker frequency* (CFF) is reached, when the light appears continuous for higher frequencies. Figure 5.4 shows how the CFF changes with brightness. Note that the projection rate of film at 48 fps and the field rate of European television at 50 fields per second are marginal with bright images.

The contrast sensitivity of the eye is defined as the smallest brightness difference which is visible. In fact the contrast sensitivity is not constant, but increases proportionally to brightness. Thus, whatever the brightness of an object, if that brightness changes by about 1% it will be equally detectable.

The true brightness of a television picture can be affected by electrical noise on the video signal. As contrast sensitivity is proportional to brightness, noise is more visible in dark picture areas than in bright areas. In practice the gamma characteristic of the CRT is put to good use in making video noise less visible. Instead of having linear video signals which are subjected to an inverse gamma

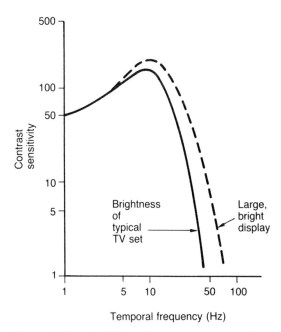

Figure 5.4 The critical flicker frequency is not constant but rises as brightness increases.

function immediately prior to driving the CRT, the inverse gamma correction is performed at the camera. In this way the video signal is non-linear for most of its journey.

Figure 5.5 shows a reverse gamma function. As a true power function requires infinite gain near black, a linear segment is substituted. It will be seen that contrast variations near black result in larger signal amplitude than variations

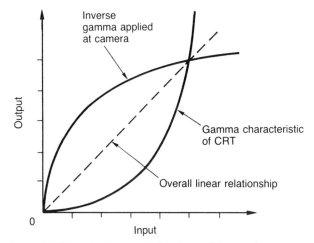

Figure 5.5 Video signals to be displayed on a CRT are subject to a reverse gamma function as shown here. In conjunction with the gamma of the CRT the overall response is made more linear.

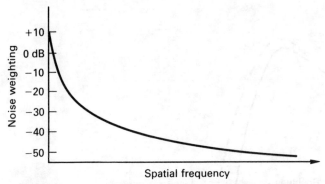

Figure 5.6 The sensitivity of the eye to noise is greatest at low frequencies and drops rapidly with increasing frequency. This can be used to mask quantizing noise caused by the compression process.

near white. The result is that noise picked up by the video signal has less effect on dark areas than bright areas. After the gamma of the CRT has acted, noise near black is compressed with respect to noise near white. Thus a video transmission system using gamma correction at source has a better perceived noise level than if the gamma correction is performed near the display.

In practice the system is not rendered perfectly linear by gamma correction and a slight overall exponential effect is usually retained in order further to reduce the effect of noise in darker parts of the picture. A gamma correction factor of 0.45 may be used to achieve this effect. If another type of display is to be used with signals designed for CRTs, then it is quite likely that some gamma conversion will be required at the display.

Sensitivity to noise is also a function of spatial frequency. Figure 5.6 shows that the sensitivity of the eye to noise falls with frequency from a maximum at zero. Thus it is vital that the average brightness should be correctly conveyed by a compression system, whereas higher spatial frequencies can be subject to more requantizing noise. Transforming the image into the frequency domain allows this characteristic to be explored.

5.2 Colour vision

Colour vision is due to the cones on the retina, which occur in three different types, responding to different colours. Figure 5.7(a) shows that human vision is restricted to range of light wavelengths from 400 to 700 nm. Shorter wavelengths are called ultra-violet and longer wavelengths are called infra-red. Note that the response is not uniform, but peaks in the area of green. The response to blue is very poor and makes a nonsense of the use of blue lights on emergency vehicles.

Figure 5.7(b) shows an approximate response for each of the three types of cone. If light of a single wavelength is observed, the relative responses of the three sensors allow us to discern what we call the colour of the light. Note that at both ends of the visible spectrum there are areas in which only one receptor responds; all colours in those areas look the same. There is a great deal of variation in receptor response from one individual to the next and the curves used

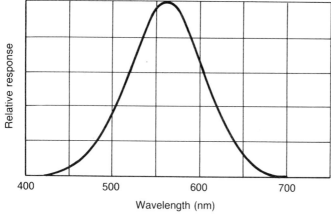

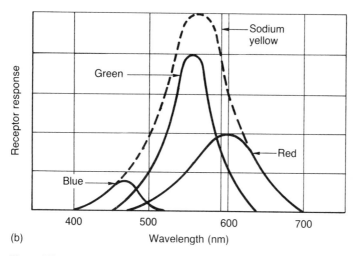

Figure 5.7 (a) The response of the eye to warious wavelengths. There is a pronounced peak in the green region. (b) The different types of cone in the eye have the approximate responses shown here which allow colour vision.

in television are the average of a great many tests. In a surprising number of people the single receptor zones are extended and discrimination between, for example, red and orange is difficult.

The triple receptor characteristic of the eye is extremely fortunate as it means that we can generate a range of colours by adding together light sources having just three different wavelengths in various proportions. This process is known as *additive colour matching* and should be clearly distinguished from the subtractive colour matching which occurs with paints and inks. Subtractive matching begins with white light and selectively removes parts of the spectrum by filtering. Additive matching uses coloured light sources which are combined.

5.3 Colour difference signals

An effective colour television system can be made in which only three pure or single-wavelength colours or *primaries* can be generated. The primaries need to be similar in wavelength to the peaks of the three receptor responses, but need not be identical. Figure 5.8(a) shows a rudimentary colour television system. Note that the colour camera is in fact three cameras in one, where each is fitted with a different-coloured filter. Three signals, R, G and B, must be transmitted to the display which produces three images which must be superimposed to obtain a colour picture.

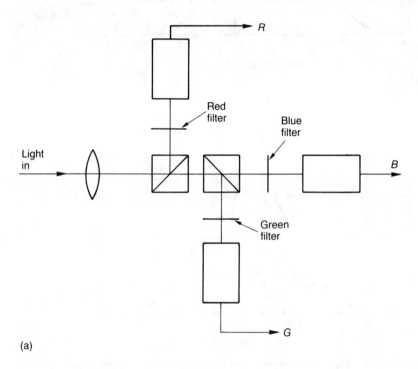

(a)

$$\begin{bmatrix} Y \\ Cr-128 \\ Cb-128 \end{bmatrix} = \begin{bmatrix} \frac{77}{256} & \frac{150}{256} & \frac{29}{256} \\ \frac{131}{256} & \frac{-110}{256} & \frac{-21}{256} \\ \frac{-44}{256} & \frac{-67}{256} & \frac{131}{256} \end{bmatrix} \begin{bmatrix} R \\ G \\ B \end{bmatrix}$$

(b)

Figure 5.8 (a) A simple colour system uses three primaries and transmits a complete picture for each. This is incompatible with monochrome and uses too much bandwidth. Practical systems use colour difference and luminance signals which are obtained by a weighted calculation as shown in (b).

A monochrome camera produces a single luminance signal Y whereas a colour camera produces three signals, or *components*, R, G and B, which are essentially monochrome video signals representing an image in each primary colour. RGB and Y signals are incompatible, yet when colour television was introduced it was a practical necessity that it should be possible to display colour signals on a monochrome display and vice versa.

Creating or *transcoding* a luminance signal from R, G and B is relatively easy. Figure 5.7 showed the spectral response of the eye which has a peak in the green region. Green objects will produce a larger stimulus than red objects of the same brightness, with blue objects producing the least stimulus. A luminance signal can be obtained by adding R, G and B together, not in equal amounts, but in a sum which is *weighted* by the relative response of the eye. Thus:

$$Y = 0.3R + 0.59G + 0.11B$$

If Y is derived in this way, a monochrome display will show nearly the same result as if a monochrome camera had been used in the first place. The results are not identical because of the non-linearities introduced by gamma correction.

As colour pictures require three signals, it should be possible to send Y and two other signals which a colour display could arithmetically convert back to R, G and B. There are two important factors which restrict the form which the other two signals may take. One is to achieve reverse compatibility. If the source is a monochrome camera, it can only produce Y and the other two signals will be completely absent. A colour display should be able to operate on the Y signal only and show a monochrome picture. The other is the requirement to conserve bandwidth for economic reasons.

These requirements are met by sending two *colour difference signals* along with Y. There are three possible colour difference signals, $R-Y$, $B-Y$ and $G-Y$. As the green signal makes the greatest contribution to Y, then the amplitude of $G-Y$ would be the smallest and would be most susceptible to noise. Thus $R-Y$ and $B-Y$ are used in practice, as Figure 5.8(b) shows.

R and B are readily obtained by adding Y to the two colour difference signals. G is obtained by rearranging the expression for Y above such that:

$$G = \frac{Y - 0.3R - 0.11B}{0.59}$$

If a colour CRT is being driven, it is possible to apply inverted luminance to the cathodes and the $R-Y$ and $B-Y$ signals directly to two of the grids so that the tube performs some of the matrixing. It is then only necessary to obtain $G-Y$ for the third grid, using the expression:

$$G - Y = 0.51(R-Y) - 0.186(B-Y)$$

If a monochrome source having only a Y output is supplied to a colour display, $R-Y$ and $B-Y$ will be zero. It is reasonably obvious that if there are no colour difference signals the colour signals cannot be different from one another and $R = G = B$. As a result the colour display can only produce a neutral picture.

The use of colour difference signals is essential for compatibility in both directions between colour and monochrome, but it has a further advantage which follows from the way in which the eye works. In order to produce the highest resolution in the fovea, the eye will use signals from all types of cone, regardless of colour. In order to determine colour the stimuli from three cones must be

136 Video compression

compared. There is evidence that the nervous system uses some form of colour difference processing to make this possible. As a result the acuity of the human eye is only available in monochrome. Differences in colour cannot be resolved so well. A further factor is that the lens in the human eye is not achromatic and this means that the ends of the spectrum are not well focused. This is particularly noticeable on blue.

If the eye cannot resolve colour very well, there is no point in expending valuable bandwidth sending high-resolution colour signals. Colour difference working allows the luminance to be sent separately at a bit rate which determines the subjective sharpness of the picture. The colour difference signals can be sent with considerably reduced bit rate, as little as one-quarter that of luminance, and the human eye is unable to tell.

The overwhelming advantages obtained by using colour difference signals mean that in broadcast and production facilities *RGB* is seldom used. The outputs from the *RGB* sensors in the camera are converted directly to Y, $R-Y$ and $B-Y$ in the camera control unit and output in that form. Whilst signals such as Y, R, G and B are *unipolar* or positive only, it should be stressed that colour difference signals are *bipolar* and may meaningfully take on levels below zero volts.

5.4 Motion and resolution

The human eye resembles a CCD camera in that the retina is covered with a large number of discrete sensors which are used to build up an image. The spacing between the sensors has a similar effect on the resolution of the eye to the number of pixels in a CCD chip. However, the eye acts to a degree as if it were AC coupled so that its response to low spatial frequencies (the average brightness of a scene) falls.

The response of the eye is effectively two-dimensional as it is affected by spatial frequencies and temporal frequencies. Figure 5.9 shows the two-dimensional or spatio-temporal response of the eye. If the eye were static, a detailed object moving past it would give rise to temporal frequencies. The

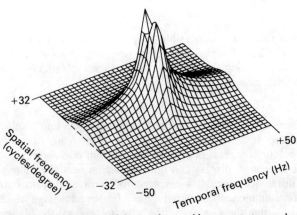

Figure 5.9 The response of the eye shown with respect to temporal and spatial frequencies. Note that even slow relative movement causes a serious loss of resolution. The eye tracks moving objects to prevent this loss.

temporal frequency is given by the detail in the object, in lines per millimetre, multiplied by the speed. Clearly a highly detailed object can reach high temporal frequencies even at slow speeds, and Figure 5.9 shows that the eye cannot respond to high temporal frequencies; a fixed eye cannot resolve detail in moving objects. The solution is that in practice the eye moves to follow objects of interest. Figure 5.10 shows the difference that eye tracking makes. In Figure 5.10(a) a detailed object moves past a fixed eye. It does not have to move very fast before the temporal frequency at a fixed point on the retina rises beyond the temporal response of the eye and there is motion blur. In Figure 5.10(b) the eye is following the moving object and as a result the temporal frequency at a fixed point on the retina is zero; the full resolution is then available because the image is stationary with respect to the eye. In real life we can see moving objects in

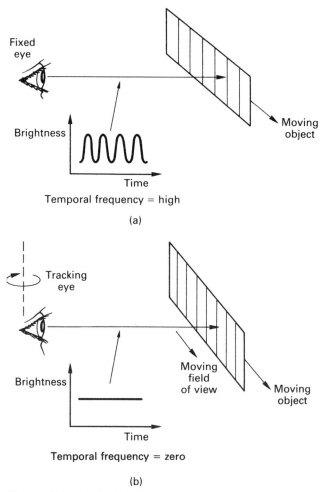

Figure 5.10 In (a) a detailed object moves past a fixed eye, causing temporal frequencies beyond the response of the eye. This is the cause of motion blur. In (b) the eye tracks the motion and the temporal frequency becomes zero. Motion blur cannot then occur.

some detail unless they move faster than the eye can follow. Exceptions are in the case of irregular motion, which the eye cannot track, or rotational motion – the eyeball cannot rotate on the optical axis!

Similar processes occur when watching television. Television viewing is affected both by the motion of the original scene with respect to the camera and by the motion of the eye with respect to the display. The situation is further complicated by the fact that television pictures are not continuous, but are sampled at the field rate.

According to sampling theory, a sampling system cannot properly convey frequencies beyond half the sampling rate. If the sampling rate is considered to be the field rate, then no temporal frequency of more than 25 or 30 Hz can be handled (12 Hz for film). With a stationary camera and scene, temporal frequencies can only result from the brightness of lighting changing, but this will not approach the limit. However, when there is relative movement between camera and scene, detailed areas develop high temporal frequencies, just as was shown in Figure 5.10 for the eye. This is because relative motion results in a given point on the camera sensor effectively scanning across the scene. The temporal frequencies generated are beyond the limit set by sampling theory, and aliasing should take place.

However, when the resultant pictures are viewed by a human eye, this aliasing is not perceived because, once more, the eye attempts to follow the motion of the scene.[1]

Figure 5.11 shows what happens when the eye follows correctly. The original scene and the retina are now stationary with respect to one another, but the camera sensor and display are both moving through the field of view. As a result

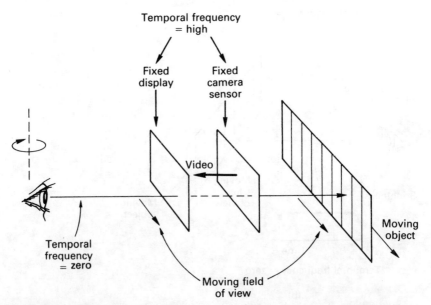

Figure 5.11 An object moves past a camera, and is tracked on a monitor by the eye. The high temporal frequencies cause aliasing in the TV signal, but these are not perceived by the tracking eye as this reduces the temporal frequency to zero. Compare with Figure 5.10.

the temporal frequency at the eye due to the object being followed is brought to zero and no aliasing is perceived by the viewer due to the field rate sampling.

Whilst the result is highly desirable, we have not actually circumvented sampling theory, because the effect only works if several assumptions are made, including the requirement for the motion to be smooth.

The viewer's impression is not quite the same as if the scene were being viewed through a piece of moving glass, because of the movement of the image relative to the camera sensor and the display. Here, temporal frequencies do exist, and the temporal aperture effect (lag) of both will reduce perceived resolution. This is the reason why shutters are sometimes fitted to CCD cameras used at sporting events. The mechanically rotating shutter allows light onto the CCD sensor for only part of the field period, thereby reducing the temporal aperture. The result is obvious from conventional photography in which one naturally uses a short exposure for moving subjects. The shuttered CCD camera effectively has an exposure control.

On the other hand a tube camera displays considerable lag and will not perform as well under these circumstances.

5.5 Applications of video compression

The data rates resulting from broadcast standard PCM video coding such as 4:2:2 are quite high, and although recorders have been developed which store the signals in this form, they will remain exclusively in the professional domain for the foreseeable future. The running costs of a professional DVTR are reasonable in the context of the overall cost of a television production, but beyond the resources of the individual. If digital video recording is to be made available to the consumer, the running cost and tape consumption will need to be reduced considerably using compression. For post-production purposes, the simplicity of PCM video is attractive because it allows manipulations with the minimum of quality loss. The signals seldom need to travel long distances and the provision of sufficient data rate is not an issue. In television, compression factors will be quite mild until all production processes are completed. A high compression factor will then be adopted for broadcasting. The increasing range of applications of digital video go beyond broadcast television and include computer graphics, videoconferencing and videophones, video-on-demand, multi-media and interactive video. All of these applications will require compression but with different quality criteria.

In some cases, transmission over long distances is required, where the cost is generally proportional to the bit rate. In new developments such as interactive video (CD-I), rapid random access is required to images on optical disk where the cost is proportional to the data stored. In both cases there is an obvious pressure to reduce the data volume, rate or both.

As PCM High Definition requires about five times the data rate of standard definition then the use of compression will be mandatory to deliver the signal to the home and mild compression will be used for production recorders. Equipment is already available to compress HDTV signals down to the data rates of D-1 and D-5 DVTRs and beyond.

The fundamentals of video compression are not new, but availability of equipment has been hampered until recently by the speed and complexity of the processing required. The cost of implementing coding techniques in VLSI circuitry has now fallen to a level where it is advantageous to use it. The large volume required for consumer applications further lowers the unit cost of such circuitry.

Compression is a flexible technology because the degree of coding complexity and the compression factor can be varied to suit the application. Video contains redundancy because typical images contain areas in which many pixels have similar values. As was seen in Chapter 1, the actual information in video is known as the entropy, which is the unpredictable or new part of the signal, and the remainder is redundancy, which is a part of the signal which is predictable. The sum of the two is the original data rate.

Once the entropy is known, a decision can be taken about how much of the entropy is to be preserved. For television production purposes all of the entropy has to be preserved in order that multi-generation processing can be performed. The degree of compression cannot be so severe that the new data rate is less than the entropy, as information must then be lost. In theory, all of the redundancy could be removed, leaving only the entropy, but this would require a perfect algorithm which would be extremely complex. In practice the compression factor will be less than this so that simpler algorithms can be used. Thus production DVTRs such as Sony's Digital Betacam and the Ampex DCT use only very mild compression of around 2:1.

For other applications, such as videoconferencing, the entropy itself will be reduced and the received images will not appear the same as the original. Value judgements must be made on the importance of these deficiencies. In non-linear off-line editors the entropy can be reduced because the images are only used to make edit decisions and are never seen elsewhere. The original recordings or films are subsequently conformed to the edit decision list at full quality.

In addition to the picture quality/entropy considerations, there are also practical matters to be considered. Codecs inevitably cause delay. In videophones and videoconferencing, delay must be kept short, whereas in video tape recording processing delay is not a serious issue. In random access applications, the decoding delay is effectively in series with the access time, whereas the encoding delay is unimportant. Long codec delays also require large amounts of memory, which raises hardware costs. In Video-CD the encoding process need not take place in real time and its cost and complexity can be spread over the large number of disks produced.

On the other hand the signal from a videophone or a television broadcast will not need to be edited. In a consumer VTR there is no requirement for multi-generation working; the editing requirement is relatively simple but there is a need for recognizable pictures in shuttle and in reverse. For production recorders and non-linear editors, editing freedom is paramount.

In computer graphics, moving images are simply regular pixel arrays which are completely updated at the picture rate. In television, 2:1 interlace is still used in which half of the lines in a frame (known as a field) are updated at a time such that the field rate is twice the frame rate.

There are thus a number of contrasting requirements for practical compression systems which clearly cannot be met by a single solution. For production purposes, *intra-coded* data reduction is used with a mild compression factor in order to allow maximum editing freedom with negligible occurrence of artifacts. Compression algorithms intended for transmission of still images in other applications such as wirephotos can be adapted for intra-coded video compression. The ISO JPEG (Joint Photographic Experts Group)[2,3] standard is such an algorithm. Wavelet transform-based codecs can also be used. The same intra-coded approach can be used for non-linear off-line editors except that higher compression factors are acceptable. Many non-linear editors allow the user to select the compression factor when the source material is compressed. Clearly if the original material is 70 mm film more data will be needed than if the material was shot on Betacam.

Where editing freedom is not a high priority, *inter-coded* data reduction allows higher compression factors. The ISO MPEG (Moving Picture Experts Group)[4] standards address these applications. MPEG-1 is a low-bit-rate system (1.5 Mbits/s) which does not recognize interlace and which significantly reduces the signal entropy. MPEG-2 is a higher bit-rate system which is flexible enough to accept interlaced or non-interlaced sources, with infrequent artifacts for the delivery of post-produced material to the consumer.

5.6 Intra-coded compression

This type of compression takes each individual picture and treats it in isolation from any other. The most common algorithms are based on the discrete cosine transform described in Section 3.6. If the input is a colour difference video signal, each component will be treated as an image and individually compressed. Thus in effect there are three parallel compression stages whose outputs are multiplexed into one data stream. In most formats the pixel spacing is greater in the colour difference signals, but for simplicity it is preferable if all three components can be processed using the same size DCT blocks. The *macroblock* has been devised to allow this. Figure 5.12 shows that a macroblock is an area of picture which contains several luminance DCT blocks for each pair of colour difference DCT blocks. The size of the macroblock is given by the transform block size and the degree of chroma subsampling.

In a 4:2:2 system the vertical colour difference sampling rate is the same as for luminance but the horizontal rate is halved. Thus a macroblock is 16 luminance pixels wide by 8 high. This screen area contains 8×8 blocks of colour difference samples and two luminance blocks of the same size side by side. In a 4:2:0 system the vertical colour difference sampling rate is also halved and so a macroblock will be 16 luminance pixels square. This will then contain four luminance blocks.

Figure 3.25 showed an example of the different coefficients of a DCT for an 8×8 pixel block, and adding these together in different proportions will give any original pixel block. The top left coefficient conveys the DC component of the block. This one will be a unipolar (positive only) value in the case of luminance and will typically be the largest value in the block as the spectrum of typical video signals is dominated by the DC component. Moving to the right the coefficients represent increasing horizontal spatial frequencies and moving downwards the coefficients represent increasing vertical spatial frequencies. The bottom right coefficient represents the highest diagonal frequencies in the block. All of these coefficients are bipolar, where the polarity indicates whether the original spatial waveform at that frequency was inverted.

In typical pictures, not all coefficients will have significant values; there will be a few dominant coefficients. The coefficients representing the higher two-dimensional spatial frequencies will often be zero or of small value in large areas

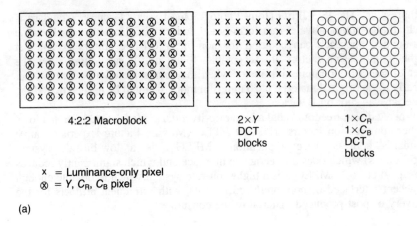

4:2:2 Macroblock   2×Y DCT blocks   1×C_R 1×C_B DCT blocks

x = Luminance-only pixel
⊗ = Y, C_R, C_B pixel

(a)

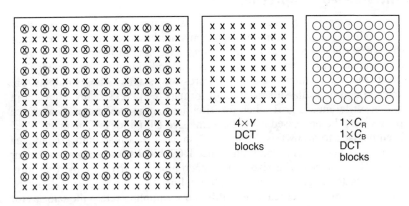

4:2:0 Macroblock   4×Y DCT blocks   1×C_R 1×C_B DCT blocks

(b)

Figure 5.12 The macroblock concept allows the same size DCT blocks to be used in chroma and luminance channels despite the different pixel spacing. Assuming an 8 × 8 DCT, for 4:2:2 working (a), the macroblock is 16 × 8 Y pixels. For 4:2:0 working (b), the macroblock must be 16 Y pixels square.

of typical video, due to motion blurring or simply plain undetailed areas before the camera. In general, the further from the top left corner the coefficient is, the smaller will be its magnitude on average. Coding gain (the technical term for reduction in the number of bits needed) is achieved by taking advantage of the zero and low-valued coefficients to cut down on the data necessary. Thus it is not the DCT which compresses the data, it is the subsequent processing. The DCT simply expresses the data in a form which makes the subsequent processing easier.

Once transformed, there are various techniques which can be used to reduce the data needed to carry the coefficients. These will be based on a knowledge of

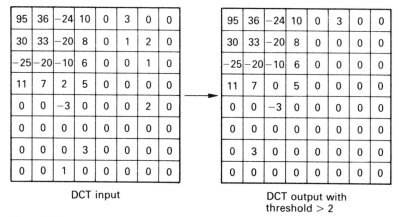

Figure 5.13 After the DCT, a simple way of compressing the data is to transmit only those coefficients which exceed a threshold.

the signal statistics and the human vision mechanism and will often be combined in practical systems.[5,6]

Possibly the simplest reduction method involves setting a threshold magnitude for coefficients. Only coefficients which exceed the threshold are transmitted, as it is assumed that smaller values make a negligible contribution to the picture block. Figure 5.13 shows the result, which is that most coefficients then have a value of zero.

As was shown in Chapter 3, coefficients are usually subject to requantizing which increases the size of the quantizing step. The smaller number of steps permits coding with fewer bits and allows coding gain.

Psycho-visual knowledge may also be used to process the coefficients. Omitting a coefficient means that the appropriate frequency component is missing from the reconstructed block. Requantizing a coefficient means that the frequency it represents is reproduced in the output with the wrong amplitude. The difference between original and reconstructed blocks is regarded as noise added to the wanted data. Figure 5.6 showed that the visibility of such noise is far from uniform. The maximum sensitivity is at DC and as a result the top left coefficient is often treated as a special case and left unchanged. It may warrant more error protection than other coefficients.

Psychovisual coding takes advantage of the falling sensitivity to noise by dividing each coefficient by a different weighting constant as a function of its frequency. Figure 5.14 shows the weighting process used in JPEG. Naturally the decoder will have a corresponding inverse weighting. This weighting process has the effect of reducing the magnitude of high-frequency coefficients disproportionately. If a coefficient threshold is used, the weighted coefficients which fail to exceed the threshold can be discarded and the result will be subjectively more acceptable than in the absence of weighting. Clearly, different weighting will be needed for colour difference data as colour is perceived differently.

In the more common case where coding gain is obtained by requantizing, the effect of weighting is that the coefficients are individually requantized with step sizes which increase with frequency. The larger step size increases the quantizing noise at high frequencies where it is less visible.

144 Video compression

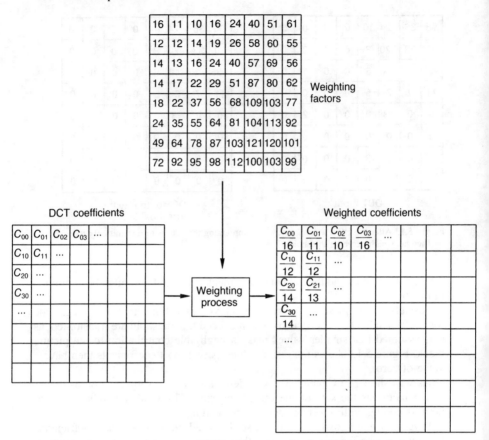

Figure 5.14 The weighting used in JPEG is based on the reduced visibility of high-frequency noise.

Knowledge of the signal statistics gained from extensive analysis of real material can be used to describe the probability of a given coefficient having a given value. This is the basis of entropy coding, in which coefficients are described not by fixed wordlength numbers, but by variable-length codes. The shorter codes are allocated to the most probable values and the longer codes to the least probable values. This allows a coding gain on typical signals. One of the best known variable-length codes is the Huffman code.[7]

The main difficulty with variable-length codes is separating the symbols when they are serialized. With fixed wordlength, the bit clock is merely divided by the wordlength to obtain a wordclock. With variable-length coding the bit stream coding must be such that the decoder can determine the boundaries between words unaided. Figure 5.15 shows an example of Huffman coding.

When serializing a coefficient block, it is normal to scan in a sequence where the largest coefficient values are scanned first. The sequence then continues such that the next coefficient is more likely to be zero than the previous one. In a square DCT block, used for non-interlaced material, a regular zig-zag scan begins in the top left corner and ends in the bottom right corner as shown in

Coefficient	Code	Number of zeros	Code
1	1	1	11
2	001	2	101
3	0111	3	011
4	00001	4	0101
5	01101	5	0011
6	011001	etc.	etc.
7	0000001		
Run-length code	010		

Figure 5.15 In Huffman coding the most probable coefficient values are allocated to the shortest codes. All zero coefficients are coded with run-length coding which counts the number of zeros.

Figure 5.16. With interlaced material the DCT block may not be square and the scan will have to be more creative. Statistical analysis of real programme material can be used to determine an optimal scan. The advantage of zig-zag scanning is that on typical material the scan often finishes with coefficients which are zero valued. Instead of coding these zeros, a unique 'end of block' symbol is transmitted instead. Prior to the last finite coefficient and the EOB symbol it is likely that some zero-value coefficients will be scanned. The coding enters a different mode whereby it simply transmits a unique prefix called a run-length prefix, followed by a code specifying the number of zeros which follow. This is also shown in Figure 5.15.

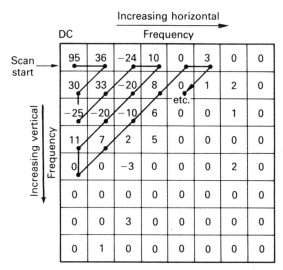

Figure 5.16 The zig-zag scan starting top left orders the coefficients in the best sequence for compression as the later ones will have smaller value.

5.7 JPEG compression

The input image in JPEG is required to be in 4:2:2 format, which means that the colour difference signals are subsampled horizontally but not vertically. As JPEG was designed primarily for photographic image transfer, the picture can be up to 65 536 pixels square. In video applications, interlaced or progressive scan signals can be handled simply by treating each field or frame as a picture. Naturally as a photographic standard JPEG has no support for audio whatsoever.

Input samples are expressed as macroblocks, also called *picture blocks*, as shown in Figure 5.12(a), which are 16 luminance pixels horizontally by 8 pixels vertically. Each picture block is then split up into blocks which are 8 pixels square. In one picture block there will be two luminance blocks and one block for each colour difference signal.

Figure 5.17(a) shows a block diagram of a JPEG compression unit. The DCT stage transforms the blocks into a form in which redundancy can be identified. Psycho-visual weighting then reduces coefficient values according to the human visual process. Zig-zag block scanning is used to place the coefficients in the order of descending priority. Coefficients are then requantized according to the compression factor required. Variable-length/run-length coding finishes the process. The resultant data are assembled into blocks and given headers for transmission.

A JPEG receiver is shown in Figure 5.17(b). The input bit stream is deserialized into symbols, and the run-length decoder reassembles the runs of zeros. The variable-length decoder then inverse-quantizes or converts back to constant wordlength coefficients. The psycho-visual weighting is reversed by a multiplication which cancels the original division. An inverse DCT then reconstructs the blocks which are assembled to form the output image. In a video application the output would be returned to a raster scan format by reading the picture buffer in the appropriate sequence.

If a visually lossless result is required, the compression factor of the system of Figure 5.17 must be a function of the input image so that all significant coefficients can be conveyed without excessively coarse quantizing. A detailed contrasty image will result in more data than a soft image containing self-similar areas. This is not a problem for single-image applications like wirephotos, because the result is that on a fixed rate link the transmission time varies slightly from one image to the next, just as it does in a fax machine.

As was shown in Section 3.10, the compression factor can be increased at will simply by raising the size of the requantizing steps. At a compression of 2:1 JPEG is nearly lossless in that many output pixels have coded values identical to the original. Multi-generation processing is quite feasible. In cascaded JPEG codecs the first requantizing is responsible for the main quality loss, after which subsequent codecs converge until each one repeats the arithmetic of the one before. Up to a factor of about 10:1 the picture remains visually lossless, but the probability of artifacts after multiple generations increases. Quality remains good on most material up to about 20:1. Beyond this the requantizing is becoming quite 'steppy' and the colour will appear posterized and may bleed across edges. When coefficients are requantized too coarsely the result is that after the inverse transform there will be an amplitude error in one or more components of spatial frequency. Certain critical amplitudes can actually be significantly increased when the requantizer rounds them up, hence the phenomenon is named high-

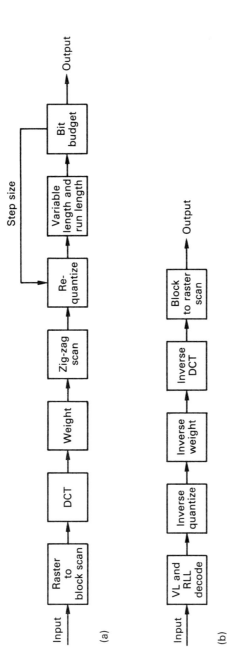

Figure 5.17 The essential stages of a JPEG codec. Note that in practice there will be three parallel channels, one for each video component.

frequency amplification. The result on decoding is that there is ringing or ripples around edges, particularly in the case of caption lettering. The phenomenon is also known as fringing or shadowing. Another result of excessive compression is that the coarse requantizing means that the video waveform at the edge of the block does not match that of adjacent blocks. The result is called 'blocking' because the DCT block boundaries become visible as a mosaic structure.

In cases where blocking and fringing artifacts are evident, the codec is really being overdriven by giving it more entropy than it can cope with. This leads to the alternative meaning of JPEG: Just Pretend Everything's Great. If a spatial pre-filtering operation is performed prior to the coder, this will remove high spatial frequencies which renders their DCT coefficients zero. The available bit rate can then be used to quantize the remaining coefficients less coarsely, thus reducing the fringing and blocking. The slight softening of the picture is subjectively preferable to the compression artifacts.

5.8 Compression in Digital Betacam

In a DVTR the picture rate is constant and variable or entropy-dependent compression is a nuisance because it demands a variable data rate. The solution is to make the compression factor constant. This is done by using feedback from the output coder which is given a bit budget. If the bit budget is exceeded, the coefficients must be requantized to larger steps. If the bit budget is under-utilized, the quantization steps are made more accurate. This is an iterative process which converges on an optimally filled output block in a variable time. This variability is undesirable in real-time hardware and instead the coefficients may be requantized by all step sizes in turn. At the end of this fixed time process the best quantization step size is then picked.

Variable quantizing results in variable noise, and it is desirable to reduce the visibility of programme-sensitive noise. One way in which this can be done is to combine a number of DCT blocks together into an entropy block. The entropy block is then given a bit budget. In this system, one DCT block having high entropy can use up the bit budget made available by other DCT blocks in the entropy block which have low entropy. Coarse quantization is not then necessary and an improvement in signal-to-noise ratio results. The chances of every DCT block in an entropy block having high entropy are not great, but can be reduced further by shuffling the DCT blocks so that an entropy block is made up from blocks which are distributed around the screen. In a DVTR this shuffle is in any case needed to break up the effects of uncorrectable errors due to dropouts.

In practice entropy blocks cannot be made arbitrarily large because only an entire block can be properly decoded owing to the use of variable-length coding. In digital VTRs the entropy block length is restricted to the track length which can be recovered in shuttle, but a useful noise advantage is still obtained.

In Digital Betacam, analog component inputs are sampled at 13.5 and 6.75 MHz. Alternatively the input may be SDI (serial digital interface) at 270 Mbits/s which is deserialized and demultiplexed to separate components. The interlaced raster scan input is first converted on a field-by-field basis to blocks which are 8 pixels wide by 4 pixels high in the luminance channel and 4 pixels by 4 in the two colour difference channels. When two fields are combined on the screen, the result is effectively an interlaced 8 × 8 luminance block with colour difference pixels having twice the horizontal luminance pixel spacing. The pixel blocks are then subject to a field shuffle. A shuffle based on individual pixels is impossible because it would raise the high-frequency content of the image and destroy the power of the data reduction process. Instead, the block shuffle helps the data reduction by making the average entropy of the image more constant. This happens because the shuffle exchanges blocks from flat areas of the image with blocks from highly detailed areas. The shuffle algorithm also has to consider the requirements of picture in shuttle. The blocking and shuffle take place when the read addresses of the input memory are permuted with respect to the write addresses.

Following the input shuffle the blocks are associated into sets of ten in each component and are then subject to the discrete cosine transform. The resulting coefficients are then subject to an iterative requantizing process followed by variable-length coding. The iteration adjusts the size of the quantizing step until the overall length of the ten coefficient sets is equal to the constant capacity of an entropy block, which is 364 bytes. Within that entropy block the amount of data representing each individual DCT blocks may vary considerably, but the overall block size stays the same.

The DCT process results in coefficients whose wordlength exceeds the input wordlength. As a result it does not matter if the input wordlength is 8 bits or 10 bits; the requantizer simply adapts to make the output data rate constant. Thus the compression is greater with 10 bit input, corresponding to about 2.4 to 1.

5.9 Inter-coded compression

High compression factors are easier to obtain if advantage can be taken of redundancy between successive images. This can be done if longer coding delays and restricted editing can be tolerated. In differential coding (DPCM), only the difference between images need be sent. Figure 5.18(a) shows that in a non-interlaced system, DPCM coding is obtained simply by inserting a one-picture period delay into the system so that a given pixel in the previous or *reference* picture is subtracted from the same pixel in the current picture to obtain a pixel difference for transmission. Clearly with a still or slow-moving picture, successive images will be similar and so the difference values calculated will be small. They can be transmitted with short wordlengths to obtain a coding gain.

In an interlaced system the pixels in one field do not coincide spatially with pixels in the previous field. It is necessary to interpolate the previous field vertically to produce a reference field as shown in Figure 5.18(b) before difference data can be calculated. The decoder will need a similar interpolator in order to recreate the interlaced output. In practice the interpolator does not need to be a high-quality device. Ordinarily, vertical image interpolation would require a complex multi-point FIR filter. In the case of a compression system, the interpolator is within the difference data loop and so simple linear interpolation between adjacent pixels can be used with a corresponding saving in complexity and cost. The data sent are the difference between the reference field and the present field. The decoder adds the difference data to the reference field to obtain an output field. If the encoder and the decoder contain identical linear interpolators they will both make the same approximations in the calculation of the reference field which will cancel out. Thus there need be no loss of quality, but the increased difference data reduce the compression factor obtained.

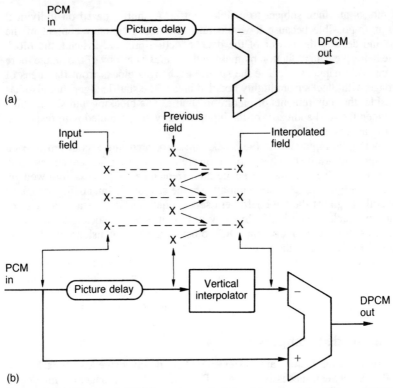

Figure 5.18 (a) In a simple DPCM system, a one-picture delay allows the difference between pixels in the same position on the screen to be computed. (b) In an interlaced system an interpolator will be necessary to create a picture having pixels in the same place as in the next field.

The difference values form a two-dimensional array which can be transform-coded to obtain a further coding gain. Too much spatial redundancy should not be expected in a difference image. The statistics of difference images are not the same as for proper images. The differentiation process results in emphasis of high spatial frequencies in the presence of motion. When requantizing the coefficients it should be borne in mind that noise in a difference image will also be perceived differently from noise in a proper image. In practice, movement reduces the similarity between successive images and the difference data increase. As a result the data flow in DPCM will rise and fall with picture content and buffering will be needed to average out the data rate.

One way of restoring the coding gain is to use motion compensation. If the motion of a part of the image from picture to picture is known, the encoder can use the motion vector to select pixel data from the *reference* image and move it in order to create a *predicted* image which is then compared with the current image. The image is broken up into macroblocks which in DCT-based systems will be integer multiples of the size of the DCT block. In inter-coding, the definition of a macroblock is that it is the area of a picture to which one motion vector applies. Figure 5.19 shows the example of CCIR Rec. 723 coding in which

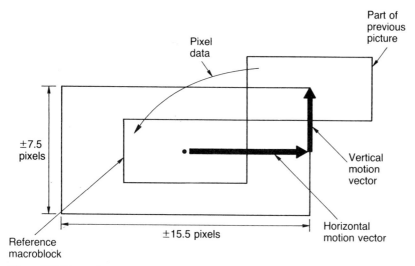

Figure 5.19 The motion vector relates to the predominant motion of an entire macroblock. In CCIR Rec. 723 a block of data can be fetched to a predicted macroblock from anywhere over a range of ±15.5 × ±7.5 pixels.

a macroblock consists of two DCT blocks side by side. The motion vector has horizontal and vertical components. In Rec. 723 the horizontal component is a 6 bit two's complement number whereas the vertical component is only represented by five bits since there is less vertical motion in typical material. The position of the macroblocks is fixed with respect to the display and so the motion compensation mechanism must work by fetching pixel values from elsewhere in the reference picture *to* the macroblock. If the motion estimation is reasonably accurate, the differences between the predicted block and the current input will be smaller than in a simple differential coder. If the motion vectors are transmitted along with the difference data, the decoder can use the vectors to produce its own predicted macroblocks to which the difference data are added. The difference between predicted and input images can be treated as an image itself and can be compressed by a DCT-based system as described above. Note that in differential coding even the DC coefficient becomes a bipolar quantity. The size of the macroblock is a compromise. If the blocks are small they can readily follow complex motions, achieving accurate prediction and hence the best coding gain. However, this gain is offset by the need to transmit more vectors. Halving the area of the macroblock doubles the number of vectors.

Shifting the contents of the reference image memory to produce the predicted macroblock is easy if the motion vectors specify integer pixel shifts. The motion vectors are simply added to the addresses used to read the reference memory. However, real motions do not respect such a requirement, and can take place to subpixel accuracy. Thus real motion vectors have an integer part which modifies the read address and a fractional part which controls an interpolator to calculate values between pixels. Once again the interpolator is within the coding loop and so need not be highly accurate as the interpolation errors are taken out in the difference data.

Figure 5.20 shows a simple motion-compensated system. Incoming video passes in parallel to the motion estimator and the line-scan to block-scan converter. The motion estimator compares the incoming picture with the previous one in the picture store in order to measure motion, and sends the motion vectors to the motion compensation unit which shifts pixels in the reference picture store output to their predicted positions in the new picture. This predicted picture is subtracted from the input picture in order to obtain the picture difference or prediction error. The result of this process is that the temporal or inter-picture redundancy and the entropy have been separated. The redundancy is the predictable part of the image as anything which is predictable carries no information. The prediction error is the entropy, the part which could not be predicted. The picture difference is then processed to remove spatial redundancy. This is done with a combination of DCT, weighting and quantizing as was explained in Section 5.6. The spatially reduced picture difference is multiplexed with the motion vectors in order to produce the system output.

It will be seen from Figure 5.20(a) that there is also a local decoder which consists of an inverse quantizer, inverse weighting stage and an inverse DCT. Adding the locally decoded prediction error (image difference) to the predicted picture must result in the original picture (plus quantizing noise) which updates the picture store. Figure 5.20(b) shows the decoder. The reference picture store output is shifted by the transmitted motion vectors and the result is the same predicted picture as was produced in the encoder. The decoded picture error is added to the predicted picture and a reproduction of the original picture results.

The motion estimation process needs to be simple and economical and it does not need to be very accurate as the motion compensation system is inside the error loop. If a motion vector is slightly incorrect, the greater picture difference data will correct for it provided there is sufficient bandwidth in the channel. If excessive compression is used, this mechanism will not function properly. The motion compensators in CCIR Rec. 723 and in MPEG work to an accuracy of half a pixel. Block matching is generally adequate for such a motion estimation application and is widely used.

5.10 Error propagation

Whilst a system based purely on inter-coding redundancy will work, it has some weaknesses. One of these is the situation where a moving object reveals background area which was concealed in the previous image. This area cannot be predicted and results in a large error. Data from such an area might just as well be intra-coded and this might be more efficient. Another problem is that the current picture is based on the history of many previous pictures. Editing will be impossible and the system will take some time to produce a picture from a cold start such as a channel switch in a receiver. Producing a picture in shuttle in a recorder is also impossible. If a transmission error occurs, the result will propagate through a large number of future pictures. A cut edit results in potentially every pixel changing value.

In practical systems these problems may be handled by switching between intra-coded, inter-coded and motion-compensated modes. There are a number of ways in which this can be done. One possibility is to transmit a signal structure in which periodically there are *I* pictures which are entirely intra-coded. Pictures between are coded using motion-compensated prediction and so are called *P*

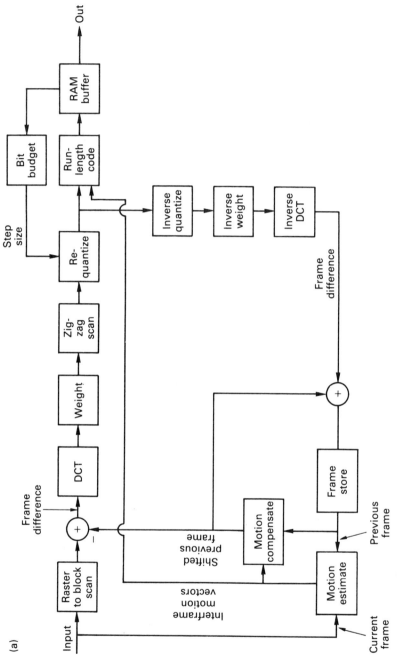

Figure 5.20 An interfield motion-compensated coder (a) and the corresponding decoder (b). See text for details.

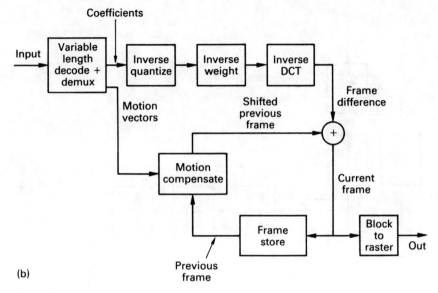

(b)

Figure 5.20 (continued).

pictures. Following a cut edit on the source material the coder could switch to an *I* picture. The *I* pictures act as 'anchors' where the picture is known absolutely and so errors cannot propagate beyond them. This is shown in Figure 5.21. Additionally it is possible to edit the compressed bit stream by switching just before the beginning of an *I* picture. *I* pictures typically contain two or three times as much data as *P* pictures.

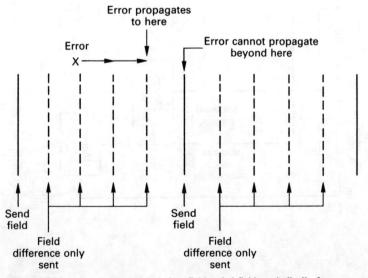

Figure 5.21 It is beneficial to include intrafield coded fields periodically for some applications.

Switching between intra- and inter-coding causes the noise level and noise structure to change. If this is done on a picture-by-picture basis, the *I* pictures contain only the requantization noise of the intra-coding whereas the *P* pictures are formed by adding noisy difference data to a noisy *I* picture and so are noisier. At high compression factors this noise pumping is visible.

If editing or picture in shuttle is not a requirement it may be better to switch encoding mode on an individual macroblock basis as required. Within the boundary of a large moving object, motion-compensated coding works best. At the boundary of a moving object where the background is being obscured and revealed there will be substantial differences from one picture to the next and so motion compensation is ineffective. In this case better results will be obtained by intra-coding the macroblock. In addition, a certain proportion of macroblocks may be forced into intra-coded mode in every picture, whether the picture content warrants this or not. The position of these blocks moves around the picture so that within a certain time period, perhaps a second, the entire picture area has been given an absolute reference. This process is known as *refreshing*, and helps to cut down the effects of error propagation.

5.11 CCIR Rec. 723 compression

CCIR Rec. 723[8] is a recommended compression scheme for contribution quality standard definition component video which produces visually lossless results at 30–45 Mbits/s: a compression of between about 5.5:1 and 4:1. There is provision for digital audio in the multiplex. The decoded signal is suitable for simple editing and mixing but multiple generation post-production and effects should be avoided. In order to cope with entropy variation a buffer of nearly 1.6 Mbits is provided at the transmitter and the receiver. The requantizing process is modified by the fullness of the buffer. Input can be 525 or 625 line interlaced video expressed as the 8 bit 4:2:2 coding level of Rec. 601 which may be input by serial or parallel interfaces to Rec. 656. Fields of 720×248 or 720×288 pixels are first divided into macroblocks. The macroblock dimensions were shown in Figure 5.12(a). Luminance and colour difference sample values are converted to two's complement coding prior to the transform process.

Three coding modes are supported on a macroblock basis: intra-field, inter-field and motion-compensated inter-frame. In inter-field mode, macroblocks are coded without reference to any other field. Inter-field mode is a form of DPCM where the current field is transmitted as the difference between the current field and the previous field. Owing to the use of interlace the previous field has to be vertically interpolated to produce a reference field. In inter-frame mode a field is coded by producing a reference field from a field of the same type in the previous frame using motion compensation. It is possible to code a macroblock in all three modes prior to selection of the appropriate mode for transmission. In the case of a cut edit or background revealed by motion, an intra-field mode may be best. In the case of little or no motion the inter-field mode (which is not motion compensated) will give the best results. In the case of rapid motion the inter-frame mode will be selected. In the two inter-modes the difference data must be within the range -128 to $+127$. If this is not achieved the mode will be forced to change. Motion vectors are calculated to half-pixel accuracy over a range of ± 15.5 pixels horizontally and ± 7.5 lines vertically. The type of motion estimation is not specified.

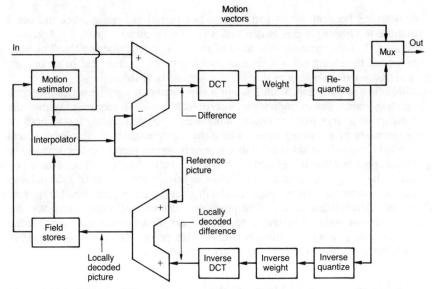

Figure 5.22 A CCIR Rec. 723 contribution quality coder. Field memories are provided to allow inter-field and inter-frame coding. These coding modes may be performed in parallel so that the choice of coding may be made later on a macroblock basis.

Figure 5.22 shows a block diagram of the coder. The motion estimator and the field memories can be seen on the left along with the interpolators necessary for handling interlace and half-pixel motion vectors. The samples or the sample differences are subject to an 8×8 DCT which outputs 12 bit two's complement coefficients. These coefficients undergo a scaling process based on their visibility and on the buffer occupancy. If the buffer approaches overflow, high spatial frequency coefficients are increasingly likely to be omitted by multiplying by zero in the scaler, thus reducing the data rate and allowing the buffer to empty. Following the scaler, coefficients are subject to a requantizer which has a piecewise linear transfer characteristic shown in Figure 5.23. Along with each macroblock of coefficients are sent the coding mode, the motion vectors (if required) and the parameters used to calculate the scale factors so that the decoder can reverse the scaling process. Finally a variable-length coding scheme

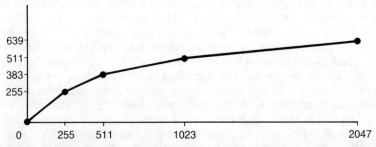

Figure 5.23 The requantizer law of CCIR Rec. 723 is non-linear on a piecewise basis and effectively has four requantizing step sizes.

5.12 Introduction to MPEG coding

In MPEG coding, an additional type of picture, known as a *B* picture or bidirectional picture, is used to allow lower bit-rate working. Figure 5.24 shows the relationship between *I*, *P* and *B* pictures. The *I* pictures are intra-coded as described in Section 5.6 to act as anchors allowing editing and channel switching. Single *I* pictures can be decoded in isolation allowing picture in shuttle in recording applications. Typically one picture in twelve (the **M** parameter) may be an *I* picture. Between *I* pictures are a number of forward-predicted *P* pictures. The distance between *P* pictures is denoted by the parameter **N**. A *P* picture is constructed in the decoder by taking the previous *I* or *P* picture and shifting it according to motion vectors transmitted for each macroblock. An inverse transform is performed on the transmitted difference data which are then added. Thus *P* pictures require around one-third the data of *I* pictures. Between the *P* pictures are a number of *B* pictures, typically two. *B* pictures are reconstructed bidirectionally using data from the *I* or *P* pictures before and after. The motion vectors between picture *B*1 and the *P* picture immediately before it are computed for each macroblock and used to produce a *forward-predicted* image. At the same time the motion vectors between picture *B*1 and the next *P* picture after it are also calculated and used to produce a *backward-predicted* picture. Both the forward- and backward-predicted pictures are then compared with the actual picture one macroblock at a time in order to measure the size of the prediction error. In some cases the forward and backward prediction errors will be of the same order, but in the area of the edges of moving objects pixels will be concealed or revealed. Figure 5.25 shows that where obscuration is taking place, background pixels in a future picture will be concealed even more and a smaller prediction error will be obtained by using data from a previous picture. Where revelation is taking place at the trailing edge of a moving object a smaller prediction error will be

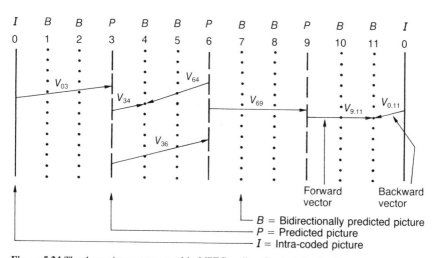

Figure 5.24 The three picture types used in MPEG coding. See text for details.

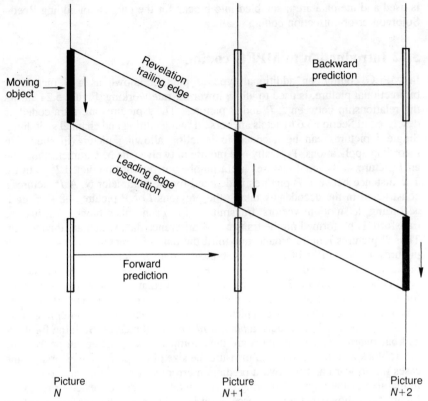

Figure 5.25 Near the edges of a moving object obscuration and revelation take place. Greater redundancy can be found if it is possible to choose whether to move forward or backward in time. Thus bidirectional coding can achieve a high compression factor.

obtained by using data from a future picture. Where there is no obscuration or revelation it is possible to reconstruct a B picture by using 50 per cent forward and 50 per cent backward prediction simultaneously. One picture is forward predicted, one picture is backward predicted, and the average of the two is used as the B prediction. This has the advantage of reducing noise, but requires more vectors to be transmitted. B pictures are not used for prediction and so if they contain errors these cannot propagate.

The coder chooses between forward, backward and 50/50 prediction for each macroblock and then transform-codes the prediction error which is transmitted along with the forward and/or reverse vectors. Choosing the prediction direction allows the B pictures to be transmitted at one-third to one-quarter the data rate of the P pictures. As both the encoder and the decoder need access to a P picture before the intervening B pictures can be encoded and decoded, the B picture data are actually transmitted *after* the P picture which follows them in the real-time sequence. This re-ordering is shown in Figure 5.26. Note that the result of using bidirectional prediction is that as much as ten pictures of memory are necessary for encoding.

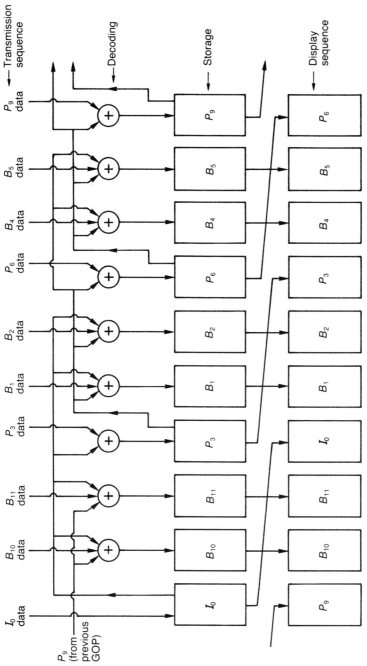

Figure 5.26 Bidirectional coding requires future pictures to be known and so pictures are transmitted out of sequence in the order in which they are *required* by the decoder.

Whilst the *I* pictures must be entirely intra-coded, the coding of the *P* and *B* pictures is flexible. Individual macroblocks in a *B* picture can be *I* or *P* coded and individual macroblocks in a *P* picture can be *I* coded. This allows the system to handle cuts and wipes which do not necessarily coincide with the *I* picture timing.

A Group of Pictures (GOP) was intended to be the smallest unit of time which could be independently decoded. GOPs are assembled into a Sequence for transmission. The Sequence contains a header which specifies the picture size, the picture rate, the bit rate and the requantizing step sizes used for each coefficient in inter- and intra-coded modes. The first GOP in a Sequence always begins with an *I* picture and in the example of Figure 5.24 ends with picture 9. Pictures 10 and 11 are considered to belong to the next GOP, illustrating that a GOP is not truly independent but retains some convolutional character.

5.13 MPEG-1 coding

MPEG-1 is a compression standard which is intended to allow moving picture reproduction with a bit rate of only 1.5 Mbits/s.[9] This is a remarkably small bit rate and was not arrived at by entropy considerations but by a requirement to shoehorn moving pictures into the data rate of a normal 'Red Book' Compact Disc player. MPEG-1 is a derivative of the H.261 videophone specification and, as might be expected, the results are acceptable on simple source material such as talking heads, computer graphics and soft material with little motion, but the quality on real pictures is grim. It is surprising that so much effort has been expended with such poor results when the high density Compact Disc (HDCD) was so close to the market. With at least four times the capacity and bit rate of a normal CD, HDCD is ideal for pre-recorded compressed video.

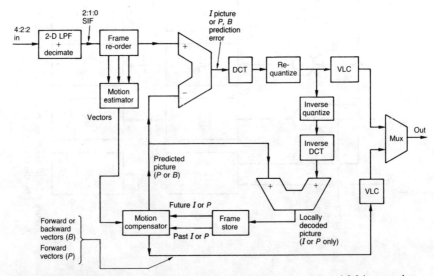

Figure 5.27 An MPEG-1 coder. Note the pre-processing necessary to convert 4:2:2 input to the 2:1:0 SIF (source input format).

The MPEG-1 coder is shown in Figure 5.27. In order to meet the desperately small bit rate allowed, MPEG-1 has not only to eliminate redundancy but also must eliminate a fair amount of entropy as well. The source entropy is reduced by subsampling in all three dimensions. If the source material is CCIR-601 component digital, 4:2:2 level, the first act of an MPEG-1 coder is to discard every other field. This halves the data rate and neatly circumvents interlace handling because the result is a progressive scan system with a frame rate of 25 or 30 Hz. Discarding every other field has the effect of halving the vertical resolution on still images, so for consistency the horizontal resolution is halved by subsampling to 360 luminance pixels per line. The chroma signals are subsampled horizontally so that there are 180 pixels per line and also subsampled vertically so that on alternate lines there are no chroma pixels. In CCIR-601 language the result is 2:1:0 coding. The result of the spatial downsampling is that each field contains $\frac{3}{8}$ of the original data. This pre-processing produces what is known as the Source Input Format (SIF). Clearly a matching post-processing stage will be required which interpolates back to the original video format.

As every other field is discarded the overall source data rate of 31.5 Mbits/s is $\frac{3}{16}$ of the original 4:2:2 signal which had 168 Mbits/s (assuming 8 bit coding). As well as reducing the resolution, discarding every other field damages the motion portrayal ability of the system and makes it the same as film. Thus the common claim that MPEG-1 offers VHS quality is simply not true.

Following the 5:1 source downsampling, MPEG-1 coding continues by assembling macroblocks which are 16×16 Y pixels. As was shown in Figure 5.12(b) this results in six DCT blocks per macroblock. Coding then proceeds according to Section 5.12 with a compression ratio of 21:1 to produce the final data rate of 1.5 Mbits/s.

The downsampled input pictures are stored in a buffer which allows them to be accessed out of sequence for bidirectional coding. At the start of a Sequence the first picture in a GOP will be intra-coded. This picture is fetched from the input buffer to the coder. In intra-mode the input subtractor is disabled and the picture blocks are directly input to the DCT. Following the transform the coefficients are weighted and requantized prior to the variable-length coder.

The requantized signal is also passed to a local inverse quantizer and inverse transform which decodes the picture in the same way as the real decoder will. This decoded picture enters the picture store system which allows the coder access to future and past reference pictures needed for bidirectional coding.

Once the I picture is coded, the first P picture is processed. Pictures in the input buffer which are to become B pictures are skipped to access the P picture. The P picture is compared with the locally decoded I picture in the motion estimator to obtain motion vectors for each macroblock. The motion compensator shifts and interpolates the I picture according to the motion vectors to produce a predicted picture. This is subtracted from the actual P picture to produce the prediction error which is transform-coded, requantized and VLC-coded. The motion vectors for each macroblock are added to form a compressed P picture. The decoder will be able to use the vectors to shift the I picture in exactly the same way to produce an identical predicted picture. If the transmitted prediction error is inverse quantized and inverse transformed it can be added to the predicted picture to create the P picture. The P data are also used in the encoder to create a locally decoded P picture from the earlier I picture. The locally decoded memory now contains an I picture and a P picture which is identical to those which the decoder will be able to create. In predictive coding, sometimes the motion-compensated prediction is exact and so the prediction error will be zero. MPEG takes advantage of this by sending a code to tell the decoder there is no error data for the macroblock concerned. The decoder can produce the picture with the motion vectors alone, hence the alternative name of Motion Predictive Educated Guesswork.

The encoder now returns to the B pictures which were skipped when the P picture was encoded. The motion estimator operates twice and calculates forward vectors from the I picture to the B picture and backward vectors from the P picture to the B picture. These are used by the motion compensator to create three predicted pictures: forward, backward and 50/50. The input B picture is subtracted from these predicted pictures in turn. The prediction which allows the prediction error to be coded with the least data is selected on a macroblock basis and the prediction error is transformed, requantized and VLC-coded. The B picture data block is assembled with macroblock coefficients and forward and/or reverse vectors. The vector data will also be VLC coded.

The output buffer is designed to even out the data flow despite entropy variations in the input pictures and data rate changes due to the differing compression factors of I, P and B pictures. If the buffer fills to a point where it is in danger of overflowing, the coefficient requantizer will increase its step size to cut the data rate. The step size is included in the data so that the decoder's inverse quantizer can track. In the case of a still picture being input the prediction circuits will give ideal performance and the data rate will dry up. In order to prevent underflow the coder produces padding data which is flagged so that the decoder can discard it.

If there is more than one B picture between the I picture and the first P picture this will be coded next. Otherwise the second P picture is created using the first P picture as a prediction basis. The process then continues for as many GOPs as are contained in the Sequence.

The MPEG-1 decoder is shown in Figure 5.28. This begins operation by identifying an I picture in the input bit stream. Compressed I pictures have no motion vectors but are transform-coded pictures. The decoder deserializes the Huffman coding to obtain coefficients, inverse-quantizes them and performs inverse DCTs to produce a picture. The adder at the output of the IDCT is disabled for I pictures so the picture passes straight to the picture buffer.

Following the I picture will be P picture data which includes motion vectors. The input demultiplexer routes the motion vectors to the motion Huffman decoder which deserializes them. The motion vectors are used by the motion compensator to shift the previous I picture in the picture buffer to produce a predicted picture identical to the one in the encoder. The incoming prediction error data are Huffman decoded, inverse-quantized and inverse-transformed to produce a prediction-error picture. This is added to the predicted picture to recreate the P picture which also passes to the picture buffer.

Next will come B picture data. The demultiplexer routes vector data to the motion compensator. Here forward or backward vectors will cause the motion compensator to fetch data from the I picture or the P picture as required to create the predicted picture. The prediction-error data are decoded and added to the predicted picture to make a B picture. Once a number of pictures are present in the buffer they can be read out in the correct order and post-processed to convert from SIF to a video standard.

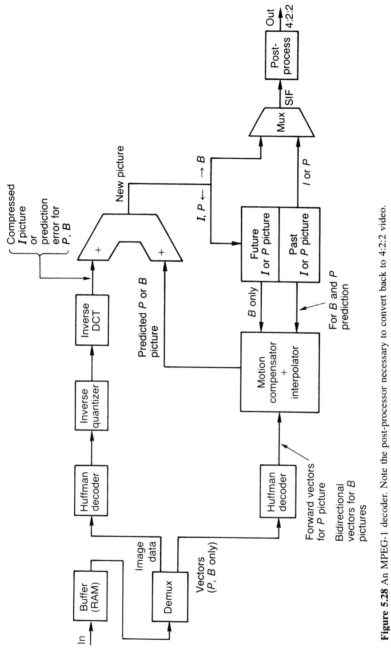

Figure 5.28 An MPEG-1 decoder. Note the post-processor necessary to convert back to 4:2:2 video.

5.14 MPEG-2 coding

MPEG-2 uses all of the coding techniques of MPEG-1 as described above, but represents a considerable improvement over MPEG-1 in that it supports larger source input formats having better resolution.[10] Variable aspect ratios are supported and it is possible to transmit a pan-scan window so that, for example, a 4:3 display can be selected by the decoder from a transmitted 16:9 picture. 4:2:0, 4:2:2 and 4:4:4 levels can be represented. The picture rate is not necessarily subsampled, and interlaced input formats can be handled. Video which has come from telecine machines can be supported. The third field of 3:2 pulldown video from a 60 Hz telecine is discarded and the remaining pairs of fields are de-interlaced to create the original film frames. These are coded as frames and the decoder recreates the 3:2 pulldown. It is possible to operate without *B* pictures if low delay is a requirement.

In fact MPEG-2 is not a single compression standard but is a standardized set of processes or tools which can be combined in various ways. One of the most powerful aspects of MPEG-2 is that it is a *layered* structure which allows the creation of *scalable* systems. A scalable system is one in which a common bit stream may be decoded to various quality levels according to the cost or complexity of the receiver. The bit stream contains *base layer* picture data which may be subsampled and/or noisy compared with the original, but which can be decoded in isolation to give a base quality picture. The data stream then contains *enhancement layer* data which can be added to the base layer picture in more complex decoders in order to improve it in some way.

Scalability may be in signal-to-noise ratio, chroma resolution, luminance resolution and temporal resolution.

In a future digital HDTV broadcasting system it may be advantageous to have a single transmitted signal which supports a variety of decoding resolutions and signal qualities. Portable receivers usually have poorer antenna location and so the base layer data could be transmitted with more error correction redundancy. A portable receiver with a small screen could discard high-frequency coefficients prior to the inverse transform.

Enhancement layers could be transmitted with less error protection as they would be received under better conditions by fixed antennas. A fixed standard definition receiver could decode an enhancement layer which is added to the base layer in order to improve the signal-to-noise ratio or reduce the visibility of compression artifacts. A high definition receiver would need to decode further enhancement layers which would add spatial resolution to the base layer. In the event of signal deterioration the presence of the strongly coded base layer would allow graceful degradation of the picture.

Figure 5.29(a) shows the principle of sub-band layering where the two-dimensional spectrum of the input HDTV picture is divided up before coding. The HDTV input is subject to low-pass filtering and downsampling in both axes to produce a normal definition picture. This is subject to data reduction as described above, and produces a first layer signal which is sufficient to drive a conventional TV set on its own.

The HDTV input is also bandpass filtered to produce a second area of the two-dimensional spectrum which becomes a 'helper' or second layer signal which when added to the normal resolution signal results in a HDTV signal once more.

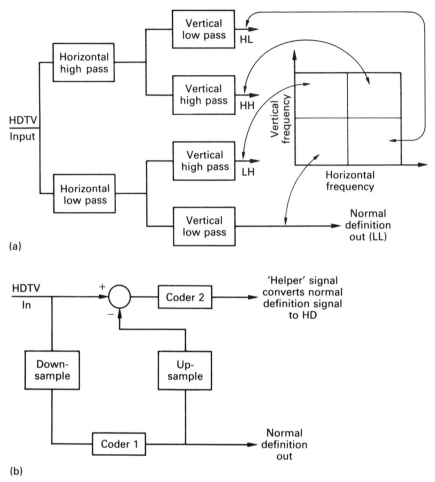

Figure 5.29 (a) A sub-band-based hierarchical coder using prefiltering. In (b) a pyramidal approach is used.

Figure 5.29(b) shows a pyramidal coding scheme in which the HDTV input is downsampled in two dimensions and fed to the base level coder which forms the normal resolution output. There is a local decoder for this signal, which is upsampled by interpolation to the HDTV line standard. This signal is then subtracted from the HDTV input to produce the helper signal. The advantage of this structure is that the downsampling filter characteristics no longer affect the coding efficiency. De-interlace is still necessary on interlaced inputs.

Temporal scaling may be performed by producing two interleaved signals each having half the picture rate of the original. In the case where the input is a progressive scan HDTV signal, it is possible to make a scalable transmission by converting the input into two parallel interlaced signals. These could be viewed individually on a SDTV display or combined to produce the original HDTV signal. Figure 5.30 shows the principle. A suitable decoder could use

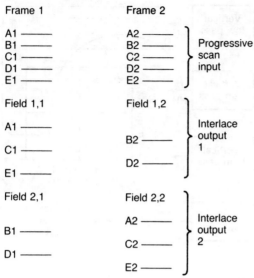

Figure 5.30 A progressive scan input may be converted into two interlaced signals which may be transmitted as different layers.

interpolation to conceal missing data in the event that one of the layers experiences a transmission error.

Figure 5.31 shows the syntax of an MPEG-2 bit stream. The largest element is the Sequence. The Sequence header contains a Sequence Start Code for synchronizing purposes. The picture size, aspect ratio, bit rate and frame rate are defined once for the entire contents of a Sequence in the Sequence header. This also contains the quantizer matrices which the encoder used so that the decoder can reverse them.

Within a Sequence are a number of Groups of Pictures (GOPs) which are the nearest thing to units of data which can be independently decoded. Typically a GOP may contain 12 frames so that a picture can be available no more than $\frac{1}{2}$ s after a channel switch. The header of the GOP contains a Group Start Code for synchronizing. The header contains the current timecode and flags which convey whether this is a closed GOP or has a broken link due to bit-stream editing.

Within a GOP are a number of Pictures, each of which contains a Picture Start Code for synchronizing. The Picture header contains the type of picture, I, P or B, and the time within the GOP where it belongs.

The Picture contains a number of Slices which in turn contain a variable number of macroblocks. Each slice contains its own synchronizing pattern, so following a transmission error, correct decoding can resume at the next slice. The slice size is variable so that it can be matched to the characteristics of the transmission channel. For example in a packet transmission system the use of a large number of slices in a packet simply wastes data capacity on surplus synchronizing patterns. However, in a recording application it might be advantageous to have frequent resynchronizing.

A 4:2:0 macroblock contains the coefficients from six DCT blocks, four Y and two chroma, whereas a 4:2:2 macroblock contains four of each. Macroblocks

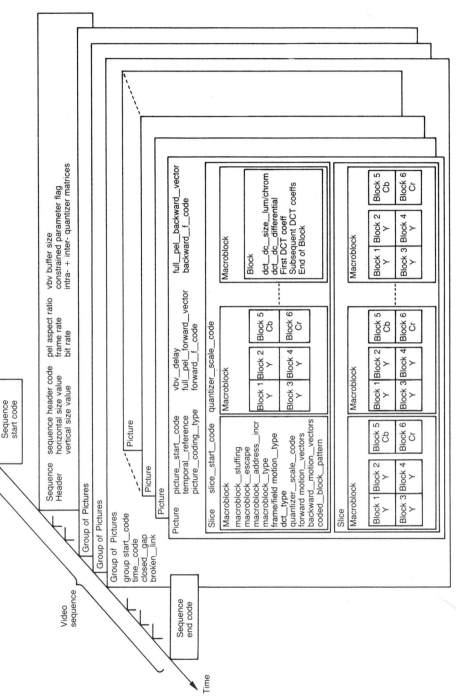

Figure 5.31 The bit stream syntax of MPEG-2. See text for details.

also contain forward and/or backward motion vectors. Although the type of picture is known, this is a default and individual macroblocks can be coded, for example, as *I* blocks in a default *P* picture. Thus each macroblock contains data describing the type of coding used as well as a flag which reflects the use of frame or field coding.

5.15 Coding artifacts

This section describes the visible results of imperfect coding. Imperfect coding may be where the coding algorithm is sub-optimal or where the compression factor in use is simply too great.

In motion-compensated systems such as MPEG, the use of periodic intra-fields means that the coding noise varies from picture to picture and this may be visible as noise pumping. Noise pumping may also be visible where the amount of motion changes. If a pan is observed, then as the pan speed increases the motion vectors may become less accurate and reduce the quality of the prediction processes. The prediction errors will get larger and will have to be more coarsely quantized. Thus the picture gets noisier as the pan accelerates and the noise reduces as the pan slows down. The same result may be apparent at the edges of a picture during zooming. The problem is worse if the picture contains fine detail. Panning on grass or trees waving in the wind taxes most coders severely. Camera-shake from a hand-held camera also increases the motion vector data and results in more noise, as does film weave.

Input video noise or film grain destroys inter-coding as there is less redundancy between pictures and the difference data become larger, requiring coarse quantizing and adding to the existing noise.

Where a codec is really fighting the quantizing may become very coarse and as a result the video level at the edge of one DCT block may not match that of its neighbour. As a result the DCT block structure becomes visible as a mosaicing or tiling effect. Coarse quantizing also causes some coefficients to be rounded up and appear larger than they should be. This causes fringing or ringing around sharp edges and extra shadowy edges which were not in the original.

Excess compression may also result in colour bleed where fringing has taken place in the chroma or where high-frequency chroma coefficients have been discarded. Graduated colour areas may reveal banding or posterizing as the colour range is restricted by requantizing. These artifacts are almost impossible to measure with conventional test gear.

Neither noise pumping nor blocking are visible on analog video recorders and so it is nonsense to liken the performance of a codec to the quality of a VCR. In fact noise pumping is extremely objectionable because, unlike steady noise, it attracts attention in peripheral vision and may result in viewing fatigue.

References

1. KELLY, D.H., Visual processing of moving stimuli. *J. Opt. Soc. America*, **2**, 216–225 (1985)
2. ISO Joint Photographic Experts Group standard JPEG-8-R8
3. WALLACE, G.K., Overview of the JPEG (ISO/CCITT) still image compression standard. ISO/JTC1/SC2/WG8 N932 (1989)
4. LE GALL, MPEG: a video compression standard for multimedia applications. *Communications of the ACM*, **34**, No. 4, 46–58 (1991)

5. CLARKE, R.J., *Transform Coding of Images*. London: Academic Press (1985)
6. NETRAVALI, A.N. and HASKELL, B.G., *Digital Pictures – Representation and Compression*. New York: Plenum Press (1988)
7. HUFFMAN, D.A., A method for the construction of minimum redundancy codes. *Proc. IRE*, **40**, 1098–1101 (1952)
8. CCIR Rec. 723 Transmission of component coded digital television signals for contribution quality applications at the third hierarchical level of CCITT recommendation G.702.
9. ISO/IEC JTC1/SC29/WG11 MPEG, International standard ISO 11172 *Coding of moving pictures and associated audio for digital storage media up to 1.5 Mbits/s* (1992)
10. ISO/IEC JTC1/SC29/WG11/602, Committee draft, *Generic coding of moving pictures and associated audio* (1992)

Index

Absolute picture data, 11
Accumulator, 57
Active filters, 29–30
ADC, 21, 30, 47, 48
Additive colour matching, 133
AES/EBU interface, 107
Aliasing, 22–7, 116, 138
Ampex DCT (digital component technology) DVTR, 3, 9, 140
Amplification, 58
Analog signals, 20
Analog video, 17
AND gates, 54, 59
Anharmonics, 42
Anti-aliasing filters, 27–30, 42, 43
Anti-image filters, 27
Aperture effect, 62, 139
Applications of compression, 1–3
apt-x100 codec, 121
Artifacts, 5, 7, 8, 15, 71, 103
   in audio, 105, 116
   in video, 18, 164
ASPEC (adaptive spectral perceptual entropy coding) System, 119, 120, 121
AT&T Bell Labs, 119
ATRAC (adaptive transform acoustic coder) coder, 125–6
Attenuation, 58
Audio compression, 8, 103–26
   applications, 107–8
   apt-x100, 121
   ATRAC coder, 125–6
   codec level calibration, 104–5
   data reduction formats, 119–20
   Dolby AC-2, 122
   floating-point coding, 110–12
   ISO Layers I, II, III, 120–1
   limits, 106–7
   masking, 103–4
   non-uniform coding, 109–10
   PASC, 122–5
   predictive coding, 112–14
   psychoacoustics, 103–4
   quality measurement, 105–6
   sub-band coding, 114–15, 117–19
   techniques, 108–9
   transform coding, 115–17
Audio muting, 39, 41
Audio sampling rate, 33–5
Audio signals, 16, 19

Backward-predicted picture, 157
Bandsplitting, 67, 98
Bandwidths, 17–20, 23, 26, 107
Bartlett window, 64
Base layer picture, 164
Basis function, 80, 86, 89
Bessel function, 66
Betts, J.A., 27
Bidirectional coding, 159, 161
Binary adding, 57
Binary codes:
   for audio, 45–50
   for component video, 50–1
Binary numbers, 39, 46, 54, 110
Binary signals, 20
Birdsinging, 42
Bit allocation, 118–24
Bit budget, 148
Bit-error-rate testing, 3, 4
Bit errors, 7, 73
Bit rate, 4
Bit rate reduction, 1
Blackman window, 66
Blocking, 148, 168
Block matching, 91–4
Broadband noise, 44
Buffering, 11
Butterflies, 81, 83–5
Butterworth configuration, 30

Capacitors, 54
Carry-in, 57
Carry-out, 57
Cascaded compression systems, 7, 15, 71, 146

171

Index

CCD cameras, 139
CCETT, 119
CCIR-601, 35, 36, 161
CCIR Rec. 723 compression, 150–2, 155–7
Channel coding, 21
Chroma, 18
Chrome cassette tape, 2
Clipping, 20, 23, 49, 57
  avoidance, 42, 113, 114, 121
Clock edge, 52
CNET, 119
Codecs, 1–2, 7, 15
  in audio, 103–7, 119, 120
  level calibration, 104–5
  in video, 140, 148, 168
Coded blocks, 118–19
Coders, 1–8, 21, 107, 113, 156
Coding artifacts, 168
Coding gain, 1, 73, 98–9, 105, 144, 150–1
  in sub-band coding, 114–18
Colour difference signals, 18, 35–7, 50–1, 134–6
Colour vision, 132–3
Compact Disc, 1, 33
Compander, 1
Companding, 109, 120
Component video, 17, 19, 38
Composite video, 17, 18, 37
Compression factor, 1, 3, 6–8
Compressor, 1, 2
Computer-aided design (CAD), 17
Computer graphics, 139
Computer simulation, 66
Constant amplitude pulse train, 23
Conversion, 21
Correlation surfaces, 97
Cosine waves, 78–86
Co-siting, 36
Critical bands, 104
Critical flicker frequency (CFF), 130
Crosstalk, 32

D-1 DVTRs, 18, 35, 139
D-5 DVTRs, 7, 18, 35, 139
DAB (digital audio broadcasting), 2, 107, 109, 119, 120
DAC, 21, 30, 41, 47
DASH, 107
Data rates, 1, 3
Data reduction, 1, 91, 106–8, 118
  formats available, 119–20
  inter-coded, 141
  intra-coded, 140
  predictive, 112
DCC (digital compact cassette), 2, 8, 109, 122
Decimation process, 71
Decoders, 1–9, 15, 21, 113, 119–21, 124, 126
De-interlaced frames, 14
DFT (discrete frequency transform), 116

Difference images, 150
Differential coding (DPCM), 149–50, 155
Difficult material, 3, 4
Digital audio, 18, 33
Digital Betacam, 3, 7, 18, 35, 87, 148–9
  Sony, 9, 140
Digital dither, 101
Digital processes, 51–2
Digital signals, 18–21
Digital video, 18, 38
Discrete cosine transform (DCT), 8, 9, 80–2, 86–9, 116, 149
  blocks, 141–2, 148–9, 151
  two-dimensional, 87, 88
Discrete Fourier transform (DFT), 73, 78, 86, 90, 93
Discrete time integrator, 57
Discrete wavelet transform, 9
Disk drives, 2, 3
Distortion, 100, 103, 112
Dither, 43–5, 101
  non-subtractive, 43
  subtractive, 43
Dolby AC-2, 114, 122
Drawbacks of compression, 7–8
DR (digital radio), 2
Dual channel sound-in-syncs (DSIS), 108
Dynamic RAMs, 55

Ear, human, 103–4, 115
EDL (edit decision list), 3, 140
Elliptic filters, 29
Enhancement layers, 164
Entropy, 4, 5, 107, 140, 144, 152, 161
Entropy blocks, 148
Error propagation, 7, 152–5
Eureka 147 project, 119
Expander, 1, 2, 3
Eye tracking, 137, 138

Fast Fourier transform (FFT), 81, 82, 84–7
Field shuffle, 149
Film-originated video compression, 13–15
Filters, 61 7
  design, 27–32
  see also under types of filters
Finite-impulse response (FIR) filters, 61–4, 67, 69, 70, 86
Flashing lights, 130
Fletcher, H., 104
Floating-point block coding, 109, 111, 122, 123
Floating-point coding, 108–9, 110–12
Folded filter, 67, 68
Forward-predicted image, 157
Fourier transforms, 8, 73–86, 89, 90, 94–5, 115
  two-dimensional, 97
Fraunhofer Society, 119

Index   173

Frequency bins, 116, 117, 119, 125–6
Frequency components, 95
Frequency domains, 73, 90
Frequency resolution, 116
Frequency response, 67
Frequency selectivity, 104
Fringing, 148, 168
Full adder, 57
Fundamentals of compression, 16–59
　aliasing, 22–7
　audio sampling rate, 33–5
　audio signals, 16
　binary adding, 57
　binary codes:
　　for audio, 45–50
　　for component video, 50–1
　conversion, 21
　digital processes, 51–2
　digital signals, 18–21
　dither, 43–5
　filter design, 27–32
　gain control by multiplication, 57–9
　logic elements, 52–4
　phase-locked loops, 37–8
　quantizing, 38–9
　quantizing error, 39–43
　reconstruction, 27
　sampling, 22–7
　sampling clock jitter, 32–3
　storage elements, 54–6
　video sampling structures, 35–7
　video signals, 16–17
　video types, 17–18

Gain control by multiplication, 57–9
Gamma correction, 132
Genlocking, 37
Gibbs' phenomenon, 64, 66
Glasberg, B., 104
Gradient matching, 91–3
Granulation, 42
Group-delay error, 61
Group of Pictures (GOP), 160, 161, 166
Group Start Code, 166

Haas effect, 104
Half adder, 57
Hamming window, 64, 66
Hanning window, 64
Hard disks, 2, 7
Harmonics, 23, 42, 73, 100, 101
HDTV, 139, 164–5
Heisenberg inequality, 76
High density compact disc (HDCD), 160
High true logic, 53
Horizontal frequencies, 87, 89
Huffman code, 5–7, 109, 121, 144, 145, 162
Human eye and vision, 128–32, 135–7
Human hearing, 103–4

Ideal quantizer, 42
IEC (International Electrotechnical
　Commission), 119
Image processing, 87
Imperfect coding, 168
Impulse response, 61–71
Infinite-impulse response (IIR) filters, 62
Information content, 4
Information rate, 4, 26
Input impulse, 62, 63
Interactive video (CD-I), 139
Inter-coded compression, 9, 10–11, 149–52,
　155
Interlace, 14, 25–7, 128, 149, 150, 155–6, 161
Interpolator, 69, 149, 151, 156
Intra-coded compression, 9, 141–5, 152, 155
Intra-coded video, 9
Intra pictures, 11
Inverse gamma correction, 131
Inverse QMF, 69
Inverse quantizers, 99, 113, 146, 152, 161
Inverse transforms, 95, 96
Inverter, 57
IRT, 119
ISO (International Standards Organization), 8,
　9, 115, 119, 122, 141
　Layers I, II, III, 120–3

Jitter, 19, 31–3
JPEG (Joint Photographic Experts Group), 9
　JPEG compression, 87, 141, 146–8

Kaiser window, 66

Latches, 54, 55, 57
Leaky predictor, 11
Lempel–Ziv–Wekh (LZW) lossless codes, 6, 7
Linearity, 39, 42
Linear phase filters, 30
Linear-phase systems, 94
Lipshitz, S.P., 102
Listening tests, 103, 105
Logic elements, 51–4
Logic gates, 53, 54
Lossless coding, 3–6
Lossy codecs, 3
Low-pass filters, 23, 27–9, 37, 62, 66, 69–71,
　73
Low true logic, 53
LSB (least significant bit), 39, 51
Luminance sampling, 36–7
Luminance signals, 17, 35, 50, 51, 135

Macroblocks, 141–2, 146, 150–1, 155–61,
　162, 166–8
Masking, 8, 103–6, 115, 118, 123
MDCT (modified discrete cosine transform),
　116, 121

Median filters, 73
Meyer, J., 30
Microphone, 16, 20, 105, 115
Midrange, 47, 49
MiniDisc, 76, 108, 109, 125, 126
Mirroring, 86, 87
Modified discrete cosine transform (MDCT), 125
Monochrome, 135
Moore, B. C. J., 103
Moore, R., 104
Morse code, 5
MOS (metal oxide semi-conductor), 54
Motion compensation, 11–14, 91, 150–3, 155, 161–2, 168
Motion detector, 14
Motion estimation techniques, 91–8
   block matching, 91–2
   gradient matching, 92–3
   phase correlation, 93–8
Motion estimator, 11, 152, 156
Motion vectors, 11–13, 151, 152, 156, 157, 162
MPEG (Moving Picture Experts Group), 9, 11, 13, 87, 119, 141, 168
   MPEG-1, 106, 108, 141, 160–3, 164
   MPEG-2, 35, 106, 108, 141, 164–8
   MPEG coding, 108, 115, 157–60, 162
MSB (most significant bit), 48, 49, 51, 57, 108, 109
MUSICAM (masking pattern adapted universal sub-band integrated coding and multiplexing), 119–21

Necessity of compression, 1
Negative logic, 53
NICAM (near instantaneously companded audio multiplex), 109
   728 stereo TV sound system, 34, 107, 111, 112, 120
Noise, 4, 6
Noise pumping, 168
Noise reduction, 4, 72, 91
Noise-to-masking ratio (NMR), 105–7
Non-ideal quantizers, 42
Non-real-time systems, 7
Non-uniform coding, 109–10
NTSC, 8, 18, 34
Nyquist rate, 33

Off-line editing, 2
Offset binary, 46–7, 50
On-line editing, 3
Optic flow axis, 11, 12, 13
OR gates, 54
Oversampling, 30, 32

PAL, 8, 18, 34
Panning, 95, 97, 168

Parallel transmission, 20
PASC (precision adaptive sub-band coding), 115, 120, 122–5
Passbands, 69
Passive filters, 29
PCM (pulse code modulation), 4, 19, 20–1, 27, 107, 114, 139
   high definition, 139
Peak-to-peak amplitude, 45, 51
Perceptive coding, 3–4, 7, 8, 18, 103, 104
Perceptual entropy, 106
Persistence of vision, 130
Phase correlation, 86, 91, 93–8
Phase-linear filter, 61
Phase linearity, 30, 94, 95
Phase-locked loops, 33, 37–8
Phase responses, 30
Philips, 119
Picture blocks, 146, 161
Picture coding, 43
Picture start code, 166
Picture store, 10
Pixels, 9, 37, 73, 140, 146–51, 155, 157, 161
   blocks, 87, 88, 91, 141
   differences, 10, 149
Polyphase filters, 71–2
Positive logic, 53
Posterization, 43
Predictability, 4, 5
Predicted image, 150
Predicted pictures, 11
Prediction errors, 152, 157–8, 161, 162, 168
Predictive coding, 8–9, 109, 112–15, 162
Predictor, 112–13
Pre-echo, 116, 117, 121, 125
Primaries, 134
Principles of compression, 4–7
Processing for compression, 61–101
   discrete cosine transform (DCT), 86–9
   filtering for video noise reduction, 72–3
   filters, 61–7
   Fourier transforms, 77–86
   motion compensation, 91
   motion estimation techniques, 91–8
   quadrature mirror filter (QMF), 67–72
   requantizing, 98–102
   transforms, 73–7
   wavelet transform, 89–91
PROMs, 57, 67
Propagation delay, 52
Pseudo-QMF filters, 71
Pseudo-video system, 33
Psychoacoustics, 103–5
Psychovisual coding, 143, 146
Pulse train, 22–3
Pure binary numbers, 48–50, 57

Q factor, 104
QMF (quadrature mirror filter), 67–72, 115, 117, 120–2, 125

Quality measurement, 105-6
Quantizing, 20-2, 38-44
  non-uniform, 121
Quantizing coefficients, 67
Quantizing error, 8, 39-44, 112, 117
Quantizing intervals, 39, 40, 42-4, 51, 100
Quantizing noise, 42, 117
Quantizing range, 42, 48-50
Quantizing steps, 99-100

Radix point, 50, 110, 111
Random access memory (RAM), 54, 107
RDAT, 107
Real-time systems, 7
Reconstruction, 27
Reconstruction filters, 27-30, 33, 34
Recursion, 72, 93
Recursive filters, 62
Redundancy, 4, 5, 11, 98-9, 140, 149
  removal, 7, 87, 91
Reed-Solomon error-correction coding, 157
Reference field, 149
Reference image, 149, 150
Refreshing, 155
Relocking, 52
Remez exchange algorithm, 66
Requantizing, 98-102, 117-18, 143, 155
  non-uniform, 108-9
Requantizing error, 105, 113
Requantizing steps, 113, 121, 146, 162
Reverse gamma function, 131
RGB system, 17, 18, 134-6
Rioul, O., 90
Ripple, 64, 66, 148
Rounding, 101, 102, 117, 168
Run-length prefix, 145

Sampling, 19, 21-7
Sampling clock, 19, 35
Sampling clock jitter, 31-3
Sampling rates, 4, 19, 63-4, 69, 107, 121, 125
  and aliasing, 23-7
  in audio, 33-5
  in video, 35-7
Scalable system, 164
Scale factors, 120-1, 124, 156
Scaling, 89
Scanning, 25
Scanning standard, 17
SDI (serial digital interface), 148
SECAM, 18
Sequences of GOPs, 161, 162, 166
Sequence Start Code, 166
Shadowing, 148
Shannon's theory, 4
Short-time Fourier transform (STFT), 74, 86
Shuffle algorithm, 149
Sidebands, 23, 69
Sigma-delta modulator, 112

Signal subtraction, 50
Signal-to-noise ratio (SNR), 19, 20, 32, 43, 107, 110, 111
Sign extension, 99
Sine waves, 78-86, 89
Slices (in GOPs), 166
Sloping waveform, 32
Sony:
  ATRAC system, 76, 125
  Digital Betacam, 9, 140
Sound pressure levels, 104-5
Source coding, 5, 21
Source input format (SIF), 161
Spatial frequencies, 9, 18, 27, 73, 136, 141, 150
Spatial luminance gradient, 92-3
Spatial sampling, 25-7
Spatial spectrum, 9
Spectra, 35, 69, 70, 73, 114
Sporting events, 91, 92
Square waves, 73-4
Stacker programs, 3
Static memories, 55
Storage elements, 51-6
Sub-band coding, 8, 109, 114-15, 117-19
Sub-band filtering, 99
Sub-band layering, 164
Subcarrier, 18
Subtractive colour matching, 133
Swedish Broadcasting Corporation, 119
Sync pulse, 50
Synthesis filter, 69

Tape speed, 2, 125
Target frequency, 79-81
TDAC, 122
Telecine machines, 13-14, 164
Television, 16, 24-5, 133-6
Television viewing, 138
Temporal frequencies, 73, 136-9
Temporal luminance gradient, 92
Temporal masking, 121
Temporal sampling, 25, 26
Thomson, 119
3:2 pulldown, 13-14
Timebase correction, 19, 32
Timebase error, 19
Time domain aliasing cancellation (TDAC), 116
Transfer functions, 39, 41
Transform-based compression, 77
Transform coding, 8, 99, 109
Transform pairs, 73, 75
Transforms, 73-7, *see also under types of transform*
Transients, 8, 116, 121
Transmission channel, 2
Transmission errors, 11, 15, 152, 166
Transversal filters, 64, 67
Transversal register, 71-2

Tweeters, 29
Two's complement system, 47–51, 57, 98–101, 151, 155

Uncertainty theory, 76

Vanderkooy, J., 102
Variable-length coding, 5, 146, 148
Vertical frequencies, 87, 89
Vetterli, M., 90
Video 8 system, 107, 109, 110
Video blanking, 39, 41
Video-CD, 1, 140
Video compression, 9, 128–68
  applications, 139–41
  CCIR Rec. 723 compression, 155–7
  coding artifacts, 168
  colour difference signals, 134–6
  colour vision, 132–3
  Digital Betacam, 148–9
  error propagation, 152–5
  human eye and vision, 128–32
  inter-coded compression, 149–52
  intra-coded compression, 141–5
  JPEG compression, 146–8
  motion and resolution, 136–9
  MPEG-1 coding, 160–3
  MPEG-2 coding, 164–8
  MPEG coding, 157–60
Videoconferencing, 139, 140
Video-on-demand, 1, 139
Videophones, 139, 140
Video sampling structures, 35–7
Video signals, 9, 16–17, 19
Video types, 17–18
Vision, 128–30
Voltage-controlled oscillator (VCO), 37

Wave filters, 71
Wavelet transforms, 8, 76, 89–91, 141
Weighting constants, 143
Window functions, 64–6, 74–6, 86, 89, 116
Wordlength, 5, 45, 59, 99–101, 123, 144
  in sub-band coding, 114, 117–18
  video, 20, 64

Zig-zag block scanning, 146
Zig-zag scanning, 145
Zwicker, E., 104

200333464